Turnaround Management for the Oil, Gas, and Process Industries

Turnaround Management for the Oil, Gas, and Process Industries

A Project Management Approach

Robert Bruce Hey

Gulf Professional Publishing
An imprint of Elsevier

ELSEVIER

Gulf Professional Publishing is an imprint of Elsevier
50 Hampshire Street, 5th Floor, Cambridge, MA 02139, United States
The Boulevard, Langford Lane, Kidlington, Oxford, OX5 1GB, United Kingdom

Library of Congress Cataloging-in-Publication Data
A catalog record for this book is available from the Library of Congress

British Library Cataloguing-in-Publication Data
A catalogue record for this book is available from the British Library

ISBN: 978-0-12-817454-8

For information on all Gulf Professional Publishing publications visit our website at
https://www.elsevier.com/books-and-journals

Publisher: Brian Romer
Senior Acquisition Editor: Katie Hammon
Editorial Project Manager: Joanna Collett
Production Project Manager: Sruthi Satheesh
Cover Designer: Mark Rogers

Typeset by TNQ Technologies

Working together
to grow libraries in
developing countries

www.elsevier.com • www.bookaid.org

To all turnaround project managers of the present and future.

Contents

Introduction

SECTION D Turnaround project closure

Conclusion

Appendices

List of figures

List of tables

List of boxes

Preface

Turnarounds are planned events, during which time production is stopped to enable the repair and inspection of any plant and equipment that cannot be accessed while the process unit is online. This book is a practical guide for those parties who manage turnarounds in any process plants that change bulk materials, either chemically or physically, into salable products. These include oil and gas, petrochemical, aluminum, cement, paper, steel, and power plants and are collectively referred to as the process industry. A turnaround could also be termed a "turnaround and inspection," a "planned shutdown," or a "planned outage."

Why another book on turnarounds?

The process industry is coming under severe pressure to cut costs. The construction of larger integrated units and the application of increasingly stringent environmental policies have been instrumental in putting a number of older operators out of business.

> *In the United Kingdom, three refineries have closed in the past 10 years leaving only six operating refineries.*[1]

Short-term cost cutting has also been proven to be highly detrimental to the industry.

> *British Petroleum's policy, under CEO John Browne (1995—2008), demonstrates the disastrous effect of cost reduction without ensuring the long-term integrity of the assets. BP managers had short assignments with incentives for cost reduction. The long-term integrity problem was passed on to the next manager, eventually with disastrous consequences in a number of cases.*[2]

The optimization of operating costs to ensure the integrity of the physical assets to produce at design capacity for the life of the investment is therefore crucial for survival of the business. Turnarounds are considered the most important events in the life of such physical assets.

The intent of this book is to influence the views of company management about the fundamental importance of capable turnaround management in relation to the health of the whole business, and to ensure that the most appropriate framework, processes, and software are applied to ensure a high probability of success. When these are applied correctly, management is empowered to manage their turnaround staff effectively to ensure that the process is on track.

The book is also aimed at turnaround project managers and senior turnaround staff who wish to upgrade their project management skills. It aligns turnaround project management with ANSI/PMI 99-001-2017, The Standard for Project Management, and ISO 21500 Guidance on project management.

Layout of the book

The book is structured in such a way that it can be used progressively through the various phases in preparation for, and in the execution of, a turnaround. There is also a more detailed discussion on standards, asset management, critical path planning, and new equipment start-up in the appendices.

Table 1 lists the primary focus areas, deliverables, audits, and approvals for each phase.

Ongoing	12-18 months prior to the turnaround	6 – 12 months prior to turnaround	3 – 6 months prior to turnaround
Phase 0: Strategic Planning	Phase 1: Initiation: Concept development	Phase 2: Work development	Phase 3: Detailed planning
Focus Areas			
1. Market conditions / swings	1. Appoint a Turnaround Project Manager and form a core team for the particular shutdown	1. Firm up all shutdown inputs – capital jobs, HAZOP studies, root cause analysis, maintenance requirements, legal requirements, SHEQ requirements, operator suggestions, process engineering requirements etc.	1. Finalized work list
2. Unit performance – process & mechanical	2. Define shutdown Charter including philosophy, objectives, budget and plan	2. Conflict resolution	2. Engineering turnovers issued
3. Legal requirements	3. Identify work input sources	3. Constructability input	3. Firm budget estimate
4. Capital planning	4. Review plant history and lessons learned (including HAZOP & root cause analysis results)	4. Scope challenge	4. Additional work procedure
5. Timing to minimize lost profit opportunities – value of downtime, cluster shutdowns, coordinate with other process plants	5. Develop work list criteria	5. Contracting plan	5. Detailed SHEQ plan including unit preparation for entry
		6. Purchasing plan	6. Critical path schedule & detailed work schedules 7. Logistics plan
Deliverables			
Annual turnaround schedule	Turnaround philosophy	Preliminary critical path schedule	Final Expenditure Request circulated for approval
Five year rolling turnaround schedule	Preliminary worklists	Final work list	Final execution plan
Forecasted turnaround budget	Order Of Magnitude (OOM) estimates	+-20% cost estimate	FEL Audit report
	Estimated turnaround duration	Improved schedule estimate	
	Turnaround preparation timetable (milestones)		
	Planning manpower forecast		
	Audit schedule		
	Management buy-in		
Phase end Audits	Phase 1 Completion Audit	Phase 2 Completion Audit	Phase 3 Completion (FEL based) Audit
IPA System Benchmarking		IPA Prospective analysis	
Phase end Approvals	Initial Budget Decision	Preliminary Budget Decision	Final Budget Decision

2 weeks to 3 months prior to the turnaround		1 month after turnaround
Phase 4: Pre-Turnaround Work	Phase 5 – Turnaround Execution	Phase 6 – Post Turnaround
Focus Areas		
1. SHEQ plan implemented	1. Process unit shutdown and inspection & repairs	1. Demobilize contractors
2. Training for contractors – orientation, fire watch, hole watch, permit receipt etc	2. Daily meetings	2. Post shutdown cleanup
3. Training for operators – fire watch, hole watch, permit issue etc	3. Daily schedule reviews & update	3. Disposal of excess material
4. MOC requirements review	4. Daily cost tracking & reporting	4. As built drawings
5. Team building	5. MOC requirements documented	5. Repair & inspection history reports
6. Shutdown sequences detailed	6. Additional work review & processing	6. Equipment and inspection databases update
7. Plant walk through	7. Repair documentation (QCPs)	7. Lessons learned and recommendations for future shutdowns
8. Temporary connections for sweetening etc	8. Pre-startup safety checks	8. Compare work done with work planned
9. Tagging – blinds, cut-ins etc	9. Startup process	9. Outstanding work for on-stream maintenance
10. Contractor mobilization	10. Cleanup unit	10. Freeze shutdown accounts
11. Cost tracking and reporting		11. Issue final cost report
		12. Issue final shutdown report
Deliverables		
Execution plan frozen and published	Turnaround execution to plan and within budget	Post-turnaround clean-up
Organisation chart published	Zero incidents	Final reports
Pre-turnaround work completed		Final audit
Field mobilization started		
FEL Audit report recommendations complete		
Training complete		
Audits		
Pre-turnaround work audit	Pre Start-up Safety Review	Post-turnaround audit
Readiness Review		Retrospective

Table 2 summarizes the 10 Knowledge Areas phase by phase.

Phase 1: initiation - concept developmt	Phase 2: work development	Phase 3: detailed planning
Timing: 12 to 18 mths prior to TA	Timing: 6 to 12 mths prior to TA	Timing: 3 to 6 mths prior to TA
1. Integration	1 Integration	1 Integration
1.0. Introduction and relevant PM processes	1.0 Introduction and relevant PM processes	1.0 Introduction and relevant PM processes
1.1. The Project Charter	1.1 Project Management Plan Development	1.1 Critical Path Model Integration
1.2. TA Philosophy/ Statement of Commitment		1.2 Inclusion of Capital Projects in a Turnaround
1.3. Project Management Plan		
2. Scope	2. Scope	2 Scope
2.0. Introduction and relevant PM processes	2.0. Introduction and relevant PM processes	2.0 Introduction and relevant PM processes
2.1. Scope Development 1	2.1. Scope Development 2	2.1 Creating the Work Breakdown Structure
2.2. Turnaround Work Order listing process	2.2. Review & Assessment Methods and Tools	2.2 Finalizing the Worklist (Validating the Scope)
	2.3. Planning Electrical Work for a Turnaround	
	2.4. Anticipating Increased Work when a Vessel is Opened	
3. Schedule	3. Schedule	3. Schedule
3.0. Introduction and relevant PM processes	3.0. Introduction and relevant PM processes	3.0 Introduction and relevant PM processes
3.1. Milestone Schedule	3.1. Scheduling	3.1 Shutdown Model
3.2 Preliminary Duration Determination	3.2. Activity Definition Process	3.2 Contingencies
	3.3. Turnaround Activity Categorization	3.3 Schedule Optimization
	3.4.Planning Prepararation for Access	3.4 Detailed Planning of Major Maintenance Projects
	3.5.Innovative Ideas for Reduction in Turnaround Duration	3.5 Freeze Date and Base Lining
4. Cost	4. Cost	4 Cost
4.0. Introduction and relevant PM processes	4.0. Introduction and relevant PM processes	4.0 Introduction and relevant PM processes
4.1. Cost Management Planning	4.1. Cost Development	4.1 Cost Estimating
4.2. Cost of a Turnaround		4.2 Cost Contingency
4.3. Cost Estimates and Budget		
5. Quality	5. Quality	5 Quality
5.0. Introduction and relevant PM processes	5.01. Introduction and relevant PM processes	5.0 Introduction and relevant PM processes
5.1. Quality Management Planning	5.1. Quality Development	5.1 Quality Control
5.2. ISO 9001	5.2. Quality Records	
5.3. Materials Quality Management		
6. Resources	6. Resources	6 Resources
6.0. Introduction and relevant PM processes	6.0. Introduction and relevant PM processes	6.0 Introduction and relevant PM processes
6.1. Project Management Appointments	6.1. Resource Levelling 1	6.1 Development of the Organizational Breakdown
6.2. Human Resource Planning	6.2. Early Ordering Activities	6.2 Detailed Planning Team
6.3. Core Team Job Functions		6.3 Resource Levelling 2
6.4. Work Scope Review Team Responsibilities		6.4 Equipment and Tools
6.5. Key Attributes of a Core Team Member		6.5 Productivity
6.6. Staff Competencies and Motivation		
6.7. Trade Resource Planning		
6.8. Booking of Specialist Services and Equipment		
7. Communication	7. Communication	7 Communication
7.0. Introduction and relevant PM processes	7.0. Introduction and relevant PM processes	7.0 Introduction and relevant PM processes
7.1. Communication Framework	7.1. Logistics	7.1 Key Procedures
7.2. Problem Solving	7.2. Review Meetings	7.2 The Plot Plan
7.3. Meeting Management: Training Example		7.3 Meetings
		7.4 Sponsor Communication
8. Risk	8. Risk	8 Risk
8.0. Introduction and relevant PM processes	8.0. Introduction and relevant PM processes	8.0 Introduction and relevant PM processes
8.1. Risk Overview	8.1. Purpose of Risk Management	8.1 Work List Contingency Review
8.2. ISO 31000 Risk Management	8.2. Project Risk Definition	8.2 Risk Register
8.3. Turnaround Risk Profile	8.3. Hazard Management	8.3 Audits & Reviews
8.4. Management Risks to be Mitigated from the Start	8.4. Audit	8.4 Incident Management
8.5. Crit and Near Crit Activity Risk Assessment		
8.6. Scope Growth Risk Assessment		
8.7. Procurement Risks		
8.8. Audits & Reviews		
9. Procurement	9. Procurement	9 Procurement
9.0. Introduction and relevant PM processes	9.0. Introduction and relevant PM processes	9.0 Introduction and relevant PM processes
9.1. Procurement Planning Overview	9.1. Constructability & Alternative Work Methods	9.1 Contract Management
9.2. Materials Management	9.2. Innovative Ideas for Reduction in Turnaround	9.2 Contractor Resource Requirements
9.3. Long Lead Items	9.3. Long Delivery Items	9.3 Model Tendering (Bidding) Process
9.4. Procurement Risk & Performance	9.4. Materials Contingency Planning	9.4 Award of Contracts and Placing of Orders
9.5. Contracting Strategies		
9.6. Division of labor		
9.7. Approaches to improve proc performance		
10. Stakeholder Management	10. Stakeholder Management	10 Stakeholder management
10.0. Introduction and relevant PM processes	10.0. Introduction and relevant PM processes	10.0 Introduction and relevant PM processes
10.1. Stakeholder Identification and Roles	10.1. Decision Executive	10.1 Final Turnaround Proposal
10.2. Getting Buy-in	10.2. Other Stakeholders	
10.3. Setting Milestones		

phase 4: pre-turnaround work	phase 5: turnaround execution	phase 6: post turnaround
Timing: 2 to 3 mths prior to TA		
1. Integration	1. Integration	1. Integration
1.0 Introduction and relevant PM processes	1.0 Introduction and relevant PM processes	1.0 Introduction and relevant PM processes
1.1 Overall Management of the pre-execution phase	1.1 Overview	1.1 Overview
	1.2 Execution Process	1.2.Detailed Analysis of Key Performance Indicators
	1.3 Change Control	
2 Scope	2 Scope	2 Scope
2.0 Introduction and relevant PM processes	2.0 Introduction and relevant PM processes	2.0 Introduction and relevant PM processes
2.1 Required Preliminary Work	2.1 Additional and New Work Validation	2.2 Statistics
	2.2 Control of Scope	2.3 Re-allocation of outstanding Work
	2.3 Scope Creep	
3. Schedule	3. Schedule	3. Schedule
3.0 Introduction and relevant PM processes	3.0 Introduction and relevant PM processes	3.0 Introduction and relevant PM processes
3.1 Pre-turnaround Activities	3.1 Progress Monitoring	3.1 Statistics
3.2 Pre-turnaround Work Completion	3.2 Monitoring and Control Activities	3.2 Schedule Reports
4 Cost	4 Cost	4 Cost
4.0 Introduction and relevant PM processes	4.0 Introduction and relevant PM processes	4,0 Introduction and relevant PM processes
4.1 Testing the Cost Management Systems	4.1 Cost Control	4.1 Statistics
	4.2 Earned Value Method	4,2 Cost Reports
5 Quality	5 Quality	5 Quality
5.0 Introduction and relevant PM processes	5.0 Introduction and relevant PM processes	5.0 Introduction and relevant PM processes
5.1 Pre-turnaround Quality Activities	5.1 Quality Control Procedure Recommendations	5.1 Quality/ Inspection Reports
5.2 Resourcing for Quality Control	5.2 Tracking Quality	5.2 The Value of Quality Records
5.3 Receipt of Materials on Site	5.3 Hold Points	
	5.4 Pressure Testing	
	5.5 Joints	
6 Resources	6 Resources	6 Resources
6.0 Introduction and relevant PM processes	6.0 Introduction and relevant PM processes	6.0 Introduction and relevant PM processes
6.1 Team Training	6.1 Daily Resource Allocation	6.1 Performance Reviews
6.2 Team Building	6.2 Productivity	6.2 Demobilization
6.3 Equipment and Tools	6.3 Fatigue Monitoring	6.3 Unit Clean-up
6.4 Resourcing the Safety Team		6.4 Return of High Value Hired Equipment
		6.5 Disposal of Excess Materials and Equipment
7 Communication	7 Communication	7 Communication
7.0 Introduction and relevant PM processes	7.0 Introduction and relevant PM processes	7.0 Introduction and relevant PM processes
7.1 Mobilization	7.1 Reporting	7.1 Demobilization Process
7.2 Final Execution Plan	7.2 Preparation for Access	7.2 Documentation
7.3 HSE Relationships	7.3 Daily Supervisor Responsibilities	7.3 Lessons Learned
7.4 Briefings and Notifications	7.4 Daily Activities of the Turnaround Project Manager	7.4 Application of Lessons Learned
	7.5 Daily Work Progress Meetings	7.5 Turnaround Report
	7.6 Conflict Resolution	7.6 Archiving Documentation
	7.7 Flawless Turnaround Awareness Program	
	7.8 Videoing Interiors of Vessels & Key Activities	
8 Risk	8 Risk	8 Risk
8.0 Introduction and relevant PM processes	8.0 Introduction and relevant PM processes	8.0 Introduction and relevant PM processes
8.1 Safety, Health, Environment and Quality Control	8.1 HSE Oversight	8.1 Post Turnaround Audit
8.2 Change Control	8.2 Addressing Known Unknowns	8.2 Risk Records and Transfer of Remaining Risks
8.3 Audit	8.3 Hand Over and Start-up	
	8.4 Comm and Start-up of New Plant and Equipment	
9 Procurement	9 Procurement	9 Procurement
9.0 Introduction and relevant PM processes	9.0 Introduction and relevant PM processes	9.0 Introduction and relevant PM processes
9.1 Pre-fabrication	9.1 Contract Management	9.1 Contractor Evaluation
9.2 Contractor Mobilization		9.2 Completion of Contractual Documentation
10 Stakeholder Management	10 Stakeholder Management	10 Stakeholder Management
10.0 Introduction and relevant PM processes	10.0 Introduction and relevant PM processes	10.0 Introduction and relevant PM processes
10.1 Final Approval to proceed with the TA Execution	10.1 Progress Reporting	10.1 Final Discussions and Project Closure
10.2 Government and Other Regulatory Authorities	10.2 Start-up Coordination	

Paragraph numbering in Chapters 3–8 is standardized from 1 to 10 in line with the 10 Knowledge Areas. There are also numerous figures, tables, examples, and case studies to support the text.

References

1. Wikipedia. https://en.wikipedia.org/wiki/List_of_oil_refineries#United_Kingdom.
2. Steffy LC. *Drowning in oil: BP and the reckless pursuit of profit*. McGrawHill; 2011.

Acknowledgments

Firstly, I wish to thank my review team members, who have been indispensable in ensuring that the book is current with regard to turnaround and asset management methodologies, and the application of suitable software: Richard McGrath in New Zealand, Antonio Conti in Belgium, and KhaiZen Foo in Malaysia.

Secondly, my gratitude to my publishers: Katie Hammon in the United States, Jo Collett in the United Kingdom and Sruthi Satheesh in India.

I am indebted to my brothers for their technical input. Douglas recently retired as CEO of an electrical switchboard manufacturing company that he started 30 years ago, and James, a marine engineer, recently retired as a forensic tribologist.

Lastly, I wish to thank Sheree, my wife and partner, for reading and editing over and over until my engineering English was readable and understandable.

Introduction

Turnarounds and the success of the company

1.0 Outline

Ideally the company needs to maximize production over the life of the physical assets. This chapter discusses why effective turnaround management is critical to the success of a company, and how a project management approach, and the correct use of appropriate sophisticated software tools, can influence this success.

1.1 Definition of a turnaround

A turnaround is often referred to as a turnaround and inspection (T&I) event since it involves inspecting and overhauling existing plant, as well as replacing worn-out plant and installing new plant. Ideally, it occurs infrequently with a number of years between each turnaround.

A clear definition, which differentiates a turnaround from an unplanned shutdown, is as follows:

> *A Turnaround is a shutdown that has been planned from more than a year in advance to ensure a run length in line with industry norms with minimal unplanned down days.*

Planned and unplanned shutdowns are generally also referred to as outages.

The turnaround skeleton or milestone plan and budget is normally included in the company business plan and budget (BP&B). Other supporting documentation is recommended as per the project charter.

Turnarounds are complex projects. Principles of project management should be applied as per the *Guide to Project Management Body of Knowledge* (PMBOK),[1] an American National Standard, or alternatively ISO 21500 Guidance on project management.[2]

Turnarounds apply people, information, and financial assets to the inspection and repair of physical assets so as to optimize the next process run length. They are critical to maximizing production over the lifetime of the asset.

Turnarounds may also include engineering work in order to enhance production, reduce environmental emissions, and increase safety.

Turnaround Management for the Oil, Gas, and Process Industries. https://doi.org/10.1016/B978-0-12-817454-8.00001-0

In summary, turnarounds are necessary to:

1) Meet statutory and/or insurance inspection requirements
2) Replenish exhausted catalysts
3) Overhaul critical rotating equipment
4) Inspect and repair the internals of critical vessels
5) Clean the internals of equipment
6) Stop leaks
7) Carry out residual life assessment
8) De-bottleneck processes by removing restraints to improve production
9) Retrofit, revamp, and modernize old plant and equipment
10) Add new plant and equipment.

1.2 Strategic view: why effective turnaround management is critical

Turnarounds directly affect the **availability** of the plant and the potential to achieve maximum production. **Days lost in a turnaround are lost forever**.

If a turnaround is not done effectively, unplanned down days increase, and the interval to the next turnaround may be below industry norms resulting in a **loss of competitive edge**.

The cost of the event, and revenue forfeited while the plant is down, are both significant.

Example 1

A calculation of the total turnaround cost (including that of idle plant based staff and the profit lost due to the plant being offline) divided by the estimated daily profit of the company will indicate the number of days that the company has to work at full production simply to finance the turnaround activity.

Example 2

A refinery with a capacity of 100,000 barrels per day and say a profit of $2 per barrel, would lose $ 200,000 for every day that the plant is down.

The application of key principles is essential to ensure **maximum production over the lifetime of the investment**. These are discussed in detail under the following headings:

1. Why efficiently managed turnarounds are integral to the long-term sustainability of the physical assets.
2. Why it is essential to have asset integrity management systems covering all aspects (operation, inspection, and maintenance) of the physical assets.
3. Why sophisticated critical path modeling for a turnaround is indispensable.
4. Why it is necessary to follow a structured project management approach to managing turnarounds.
5. When and why a turnaround project manager (TPM) should be appointed.

1.3 Why efficiently managed turnarounds are essential to the long-term sustainability of the physical assets[3]

1.3.1 The relationship between turnarounds and profit

Turnarounds play a significant role in the profitability of a company. The exclusion of upcoming outages (planned and unplanned) can be disastrous when predicting earnings per share.

Example[4]

Estimates of earnings per share for Exxon Mobil, Royal Dutch Shell, and Chevron did not take outages into account. When outages were included, Exxon posted earnings per share 27% lower than estimates, which resulted in $11 billion loss in stock within just 1 hour of trading.

Maintenance and inspection competence within a mature maintenance and inspection system are the foundations for determining good key performance indicators (KPIs) and strategies. This, together with a well-structured turnaround framework and competent turnaround management staff, ensures an optimum turnaround duration for an optimum operating cycle or run length. This in turn determines ideal production over the long term, thus maximizing long-term revenue generation. Maintenance and inspection costs come down as reliability improves with the optimization of the turnaround duration and operating cycle, thus reducing operating expense and increasing profit.

Fig. 1.1 depicts this relationship.

The maturity of established systems is discussed further in Section 1.4.3. The effect of system maturity on the increase in reliability and reduction of maintenance costs for three gas plants is shown in Box 1.3.

1.3.2 Utilization and availability

The ability of the plant to achieve **maximum sustainable capacity** relates to two key indicators: utilization and availability.

Utilization

Utilization is the answer to the question:

"What is the actual throughput measured against the design capacity or, alternatively, compared to the maximum sustainable throughput?"

Utilization is measured as a percentage of maximum sustainable throughput, determined by one of the following:

1. Design capacity
2. EPC project commissioning tests
3. Maximum sustainable production rate based on a time frame of at least 30 days without damage or premature aging of the catalyst or equipment.

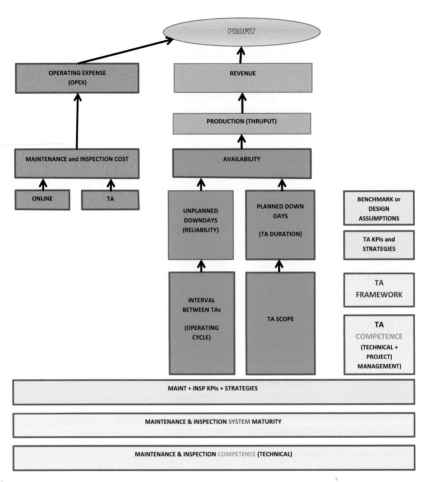

FIGURE 1.1

Turnarounds and profit relationship.

Refineries

Maximum sustainable capacity and actual throughput is based on feed flow for all refining units (barrels per day or tons per year), with the exception of the sulfur recovery unit where throughput is based on sulfur produced (tons per day).

Gas plants

Maximum sustainable capacity and actual throughput are based on production of primary products as follows:

A. LNG:
- LNG rundown capacity/production in million tons per annum.

B. Flowing/Sales Gas:
- Flowing gas capacity/production in billion cubic feet per annum.

Petrochemical plants
Maximum sustainable capacity and actual throughput are based on production of tons of primary products.

Power plants
This is based on the design capacity of the turbogenerators (MW). At times this can be exceeded, but occasionally never achieved.

Availability
Availability is determined by the following:

1. Unplanned maintenance
2. Planned maintenance
3. Consequential downtime (upstream or downstream restrictions)
4. Economic or contractual arrangements (for example, a feed gas contractual ceiling)
5. Slowdowns (normally less than 75% for more than a day)
6. Other shutdown time.

Availability is primarily dictated by maintenance downtime in days. It is split into unplanned and planned days and normally referred to as **mechanical availability (MA)**.

Unplanned down days for maintenance: These are, as the name suggests, unpredictable. These days can also be represented as **reliability** when calculated as (total days in a year − unplanned maintenance days) × 100/total days in a year.

> *Example*
>
> *A plant is shut down due to the failure of a product feed pump.*

Planned down days for maintenance: This is referred to as a planned shutdown or turnaround, which is usually planned from more than a year in advance to ensure a run length in line with industry norms with minimal unplanned down days.

Planned down days are determined in a number of different ways:

1. Actual
2. Annualized based on the last actual interval (PTAI[5] and Solomon[6])
3. Annualized based on the estimated interval if the actual intervals have been erratic
4. Annualized based on a fixed interval such as 6 years
5. Annualized based on future intervals as published in the annual BP&B.

PTAI and Solomon are benchmarking consultants used by refineries and gas plants.

Simple MA is calculated from total lost maintenance days for a particular year.

As per Solomon, annualized **MA** is generally calculated as follows:

$$MA = (\text{days in current year} - TADD - RMDD) \times 100/\text{days in current year}$$

Where:

TADD: Turnaround annualized down days taken from the last reported turnaround. Annualized days are calculated by dividing the reported days by the number of equivalent years between the last two reported turnarounds.

TADD = days down $\times$ 365.25/interval between turnarounds in days.

RMDD: Routine maintenance down days, the annualized down days for unplanned down days that are not accounted for in the turnaround data. The actual days for the previous 2 years are averaged to get an annual average.

PTAI don't average the unplanned down days for the last 2 years. They take the actual unplanned down days for the study year. PTAI benchmarks every year where Solomon benchmarks every 2 years.

The partial build of refinery availability showing the required planned shutdown days for fluidic catalytic cracking and crude units to achieve 96% availability is depicted in Box 1.1.

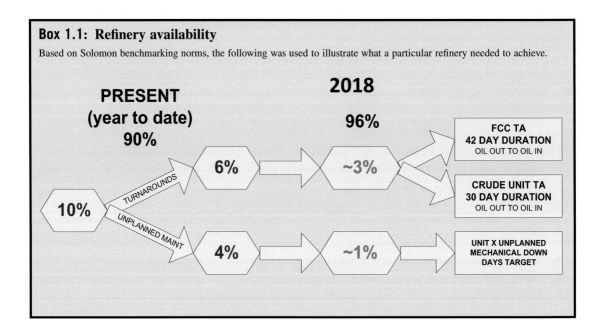

Box 1.1: Refinery availability

Based on Solomon benchmarking norms, the following was used to illustrate what a particular refinery needed to achieve.

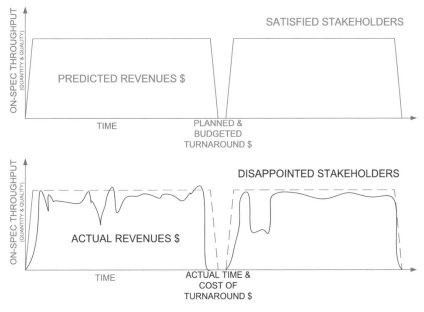

FIGURE 1.2

Revenue comparison.

The importance of availability has a direct impact on revenues as shown in Fig. 1.2.

The relationship between utilization and availability is commonly depicted as in Fig. 1.3.

The importance of utilization is illustrated in the following examples:

Example of poor utilization[7]

Nigerian National Petroleum Corp affiliates Port Harcourt Refining Co. Ltd., Warri Refining and Petrochemical Co. Ltd., and Kaduna Refining and Petrochemical Co. Ltd have a collective utilization capacity of below 20%.

Example of high utilization[4]

"Phillips 66 was one of three refiners to blow away investor expectations for the second quarter, more than doubling its earnings from a year earlier with 100 percent utilization at the company's fuel processing plants. Valero Energy Corp. and Marathon Petroleum Corp. also beat analyst's expectations."

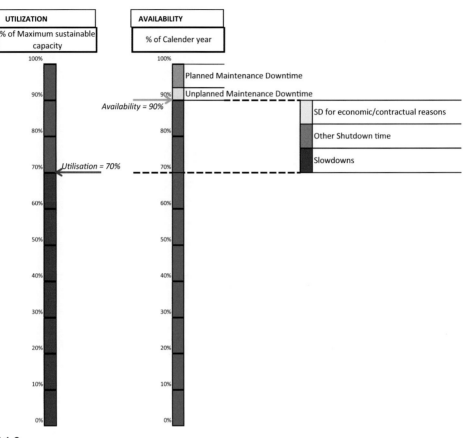

FIGURE 1.3

Relationship between utilization and availability.

1.3.3 Mechanical availability drilldown

Benchmarking can show that a target is achievable, but begs the question: How to get there?

MA is one of the most important of the KPIs. The establishment of targets for this indicator needs to be based on a long-term detailed strategy. Other discussion on MA is in Sections 1.3.2 and 2.3.

Improvement in **MA** is based on **optimizing the full operating cycle**, which includes both turnaround time and onstream time.

Fig. 1.4 depicts the **full operating cycle**.

The onstream period tends to be long (often about 5 years for refineries and gas plants), whereas the (planned) shutdown or turnaround tends to be short and intensive—planned by the hour and taking just weeks.

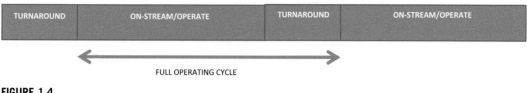

| TURNAROUND | ON-STREAM/OPERATE | TURNAROUND | ON-STREAM/OPERATE |

FULL OPERATING CYCLE

FIGURE 1.4

Full operating cycle.

As per Solomon, **annualized MA** is generally calculated as indicated in Section 1.3.2.

It is important to review the unplanned down days in the previous onstream period to identify those issues that need to be addressed in order to reduce them. Targets for **reliability** and predicted **RMDDs** can then be determined based on strategies designed to reduce unplanned down days.

Reliability = (total days in a year − unplanned maintenance days) × 100/total days in a year.

Predicted **TADDs** can then be determined from "work-shopping" a predicted run length based on Section 2.3 and modeling the next turnaround on the critical path software.

Predicted **MA** can also be ascertained and compared to the "pacesetter" value and possibly further adjusted, keeping in mind that strategies for improvement have to be realistic.

These determinations are not easy in refineries, as each type of process unit has different turnaround times, run lengths, and reliability. However, the primary process units such as crude and catalytic cracking units will dictate to the rest. The use of Solomon Associates' Profile II model to carry out "what ifs" could help in determining suitable targets.

Enhanced risk-based inspection (RBI), reliability-centered maintenance (RCM), and turnaround management are key requirements for improvement. Certification to ISO 55001 "Asset Management" would ensure continuous improvement.

"Rules of thumb" could help guide the process. Some relevant rules of thumb include:

- *Turnarounds duration*
  - *Crude unit: 28 days*
- *Turnarounds interval*
  - *Gas trains (turbines): 5.5 years*
- *Reliability*
  - *LNG mega-train design basis: 98% reliability*
    - □ *7 unplanned down days per year.*

Box 1.2 gives an example of lower-level targets rolling up to achieve long-term targets for MA and maintenance (cost) index.

Box 1.2: Refinery long-term target example

IAT: Inspection advice ticket

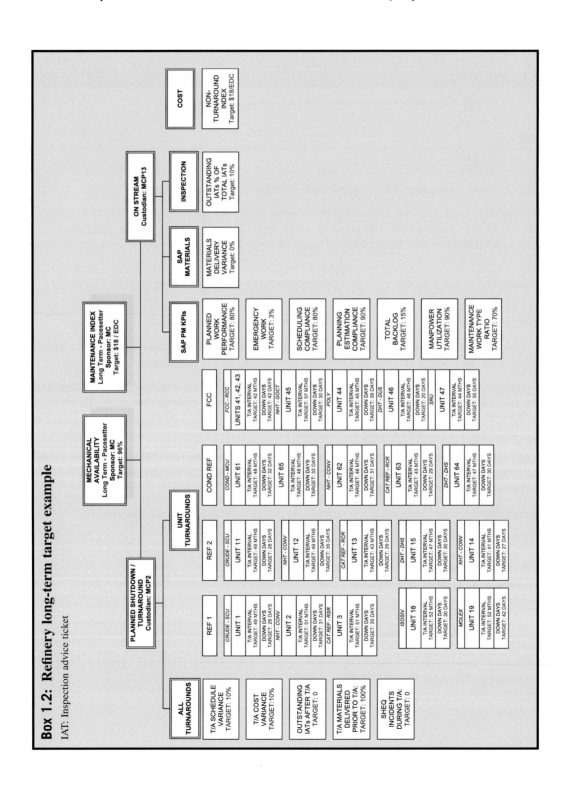

1.4 Why it is essential to have asset integrity management systems covering all aspects (operation, inspection, and maintenance) of the physical assets

1.4.1 Asset management system standards

The International Standard for Asset Management ISO 55000 is a useful tool for ensuring continuous improvement. There are three standards in the suite. They are:

ISO 55000: 2014[8]

This provides an overview of asset management, its principles and terminology, and the expected benefits of adopting asset management. It can be applied to all types of assets and by all types and sizes of organizations. It also provides the context for ISO 55001 and ISO 55002.

ISO 55001

This specifies requirements for an asset management system.

ISO 55002

This gives guidelines for the application of ISO 55001.

The above standards supersede BS PAS 55: 2008.

Benefits of adopting the ISO 55000 suite of standards

These standards enable an organization to achieve its objectives through the effective and efficient management of its assets. The application of an asset management system provides assurance that those objectives can be achieved consistently and sustainability over time.

1.4.2 Asset performance management and integrity

There is a direct link between an optimized operating cycle and the profitability of the business, and this link is critical for the survival of the business.

The integrity of the plant is tested as the boundaries of performance are pushed.

1. The key is to maximize long-term availability and to operate at design capacity.
2. The focus must be on (hydrocarbon) containment for the duration of the investment while optimizing throughput. Personnel safety and prevention of damage to the environment are integral to this.
3. Minimizing downtime (planned and unplanned) and loss of primary containment over the long term leads to maximum availability, plant integrity, and sustainability.

Fig. 1.5 depicts the relationships.

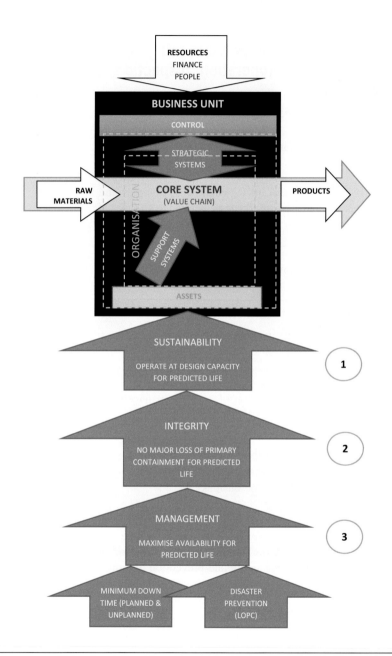

FIGURE 1.5

Performance management/integrity relationship.

1.4.3 System maturity

With each step of improvement there is a learning curve where costs may go up due to extra staffing and training before settling at a lower cost level. Similarly, reliability may also decrease as the reliability boundaries are pushed before settling at a higher reliability level.

As maintenance and inspection systems have evolved, from the implementation of computer-based maintenance management systems with associated materials modules, through the development of RBI and RCM tools, to having an integrated asset management system, maintenance costs have come down and reliability of equipment has improved.

Box 1.3 gives an example of three similar gas plants at different stages of maturity.

Support software is discussed in detail in Appendix B.

1.5 Why sophisticated critical path modeling is indispensable for complex turnarounds

1.5.1 Critical path modeling

Critical path modeling (CPM) is the most advanced and popular scheduling technique for planning, scheduling, and controlling process plant turnarounds. CPM is discussed in detail in Appendix C.

Box 1.3: System maturity − gas plants

A 2013 analysis of three similar gas plants determined how system maturity affects KPIs.

Two gas plants had implemented Meridium asset integrity software. Gas plant 1 had had the software for some time and was the pacesetter in the industry with, amongst other KPIs, the lowest maintenance and inspection costs, and the lowest staff turnover. Gas plant 2 had recently purchased Meridium and was still in the learning stages with higher staff numbers and other costs. Gas plant 3 did not have any integrated asset management, but was believed to be pursuing certification to ISO 55000 "Asset Management." It was already certified to ISO 9001, ISO 14001, and ISO 45000 within its integrated management system. The other two gas plants had not intended, at that stage, to certify to ISO 55000.

This is graphically shown here:

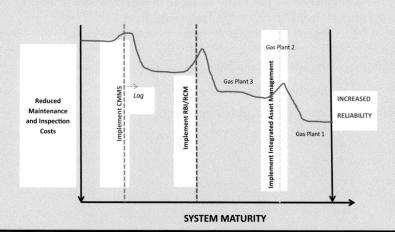

Optimum downtime is determined by modeling all activities and determining and minimizing the **longest path of activities with optimum use of resources** required for the turnaround. A software critical path model such as Oracle Primavera or Deltek Open-Plan is essential.

Analysis of the model must include thorough evaluation of the critical path and near-critical path activities. The number of near-critical path activities can affect the probability of achieving the planned completion date. If the model is well structured, probability analysis of the completion date could be undertaken. PrimaVera Risk Analysis (formerly Pertmaster) is a useful tool for this purpose. Open Plan also includes a risk function.

Organizational breakdown structure (OBS), work breakdown structure (WBS), cost breakdown structure (CBS), and resourcing must be integrated and comprise:

- OBS
  - Identifies all turnaround project supervision and management (staff and contractors)
  - Includes an OBS chart that is distributed at the start of the turnaround and posted on bulletin boards and walls so that information about who is responsible for what is visible and easy to access
  - Used to create 48 hour "look ahead" printouts that are easily available to those responsible for actions
- WBS
  - All work must be included under this hierarchy by area, discipline and system
- CBS
  - Every activity must have a cost attached
  - Every cost must relate to an activity
- Contractor breakdown structure (could be part of OBS or separate).

1.5.2 Advantages of CPM

Advantages of CPM include:

1. Models can be archived and copied for next turnaround
2. Work scope and resourcing can be produced quickly for emergency shutdowns from archived model/s and opportunity work can easily be added to the emergency work
3. Negates the need for building each turnaround model from scratch
4. Progressive improvement of the model for each successive turnaround.

It is important to have the final critical path model agreed by all interested parties involved.

Fig. 1.6 is a graphical depiction of the critical path model.

Modeling software is discussed in detail in Appendix C.

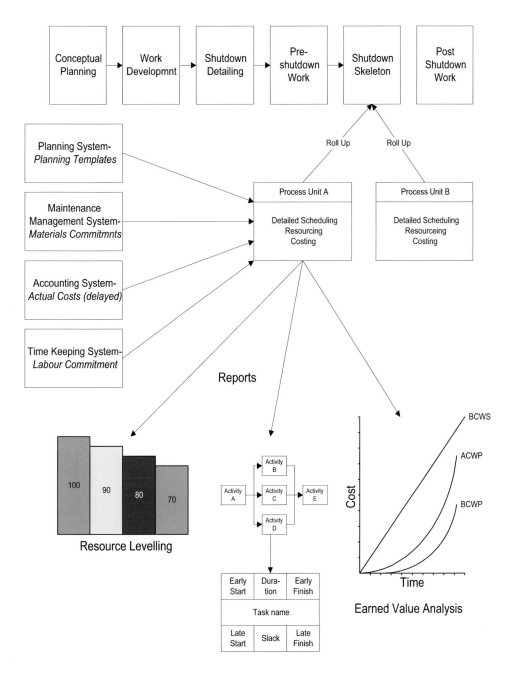

FIGURE 1.6

Critical path model overview.

1.6 Why it is necessary to follow a structured project management approach to managing turnarounds

To ensure completion on time and within budget, and to a required quality standard, a project has a high probability of success if a formal project management approach is adopted.

1.6.1 Definition of a project

A temporary endeavor undertaken to create a unique product, service, or result.

- *Temporary: Has a definite beginning and a definite end, not an ongoing effort.*
- *Unique: Doing something that has not been done before.*

Each turnaround is technically similar to previous turnarounds on that particular plant, but the time and resources would be different. The added uniqueness of each turnaround project occurs during the execution phase where progress is **measured in hours and reported daily.** Capital investment project progress is generally measured in days and reported weekly.

Turnarounds can therefore be defined as projects.

The basic principles of project management are:

- **Gates for decisions** on major development/expenditure **phases**
- **Requirements** for submission for approval
- **Review** authorities for aspects of proposal document
- **Approving** authority.

Turnarounds require a **phased** approach, with audits (**review**) and package (**requirements**) **approval** by a decision executive at each **gate**.

1.6.2 Standards

Compliance with the following international standards is common practice:

- ISO 21500 guidance on project management September 2012[2]
- A Guide to Project Management Body of Knowledge (PMBOK) ANSI 99-001-2017.[1]

These are discussed in more detail in the following paragraphs.

ISO 21500 guidance on project management September 2012

This International Standard provides guidance for project management and can be used by **any type of organization**, including public, private, or community organizations, and for **any type of project, irrespective of complexity, size, or duration.**

It provides high-level descriptions of concepts and processes that are considered good practice in project management where projects are placed in the context of programs and project portfolios.

However, this standard does not provide detailed guidance, and topics pertaining to general management are addressed only within the context of project management.

A guide to Project Management Body Of Knowledge (PMBOK) ANSI 99-001-2017
Known as A Guide to the Project Management Body of Knowledge, 6th ed. (PMBOK), it is also an American Standard (ANSI/PMI 99-01-2017) and has been translated into numerous languages.

It describes the concepts, skills, and techniques unique to the project management profession and identifies the knowledge that project managers should have to effectively manage projects.

Both ISO 21500 and PMBOK's primary focus is on balancing competing project constraints, including:

1. integration
2. scope
3. schedule
4. cost
5. quality
6. resources
7. communication
8. risk (including health, safety, and environmental issues)
9. procurement
10. stakeholder management

These will be discussed more as we progress through the phases of the turnaround. Appendix A discusses generic issues relating to compliance with ISO 21500 and PMBOK.

1.6.3 Other definitions
Project management
The application of knowledge, skills, tools, and techniques to project activities to meet project requirements.

Project stakeholders
Individuals and organizations that are actively involved in the project, or whose interests may be affected as a result of project execution or project completion.

Portfolio management
PMBOK quote:

> *"Portfolio Management refers to the centralized management of one or more portfolios, which includes identifying, prioritizing, authorizing, managing, and controlling projects, programs and other related work, **to achieve specific strategic business objectives**. Portfolio management focuses on ensuring that projects and programs are reviewed to prioritize resource allocation, and that **management of the portfolio is consistent with, and aligned to, organizational strategies.**"*

Should a company manage many turnarounds every year, then portfolio management is highly relevant. However, where there is only one turnaround every 4 years or so, then it is not necessary.

Example

Expanded portfolio management

Industry forum (including a website showing all major turnarounds in the country/region for the coming year) to coordinate turnarounds in a country or region so as to smooth turnaround specialist resources in the area.

Project management office

The project management office (PMO) is an organizational unit created to centralize and coordinate the management of projects under its domain. It is:

a. A person or number of people managing, controlling, and providing direction to temporary corporate endeavors.
b. An office, either physical or virtual, staffed by project management professionals who serve their organization's project management needs.

It also serves as an organizational center for project management excellence.

It provides project management services ranging **from** PM support functions in the form of training, software, standardized policies, and procedures, etc. **to** actual direct management and responsibility for achieving the project objective.

Many large companies have separate turnaround departments. These sometimes reflect the concept of a PMO as described in PMBOK.

Examples

- *Bahrain Petroleum Co: Shutdown and Major Maintenance Department*
- *Petronas Malaysia: Corporate Turnaround Management Department*
- *Eskom South Africa: Power Station Turnaround Group*

Project management system

The project management System:

- is the set of tools, techniques, methodologies, resources, and procedures used to manage a project;
- is a set of processes and related control functions that are consolidated and combined into a functioning, unified whole;
- aids a project manager in effectively guiding a project to completion.

Typically, one of the PMO functions would be to manage the project management system to ensure consistency in application and continuity across various projects.

1.6.4 Project integration management

Project integration management satisfies the needs and expectations of stakeholders by utilizing various tools and techniques within knowledge areas, thus ensuring success of the project.

These are:

1. **Integration**: the processes required to identify, define, combine, unify, coordinate, control, and close the various activities and processes related to the project.
2. **Scope**: the processes required to identify and define **only** the required work and deliverables.
3. **Schedule**: the processes required to both schedule project activities and to monitor progress to control the schedule.
4. **Cost**: the processes required to develop the budget and to monitor progress in order to control costs.
5. **Quality**: the processes required to plan and establish quality assurance and control.
6. **Resources**: the processes required to identify and acquire adequate project resources such as people, equipment, materials, and tools.
7. **Communication**: the processes required to plan, manage, and distribute information relevant to the project.
8. **Risk**: the processes required to identify and manage threats **and** opportunities.
9. **Procurement**: the processes required to plan and acquire products, services or results, and to manage the supplier relationship.
10. **Stakeholder**: the processes required to identify and manage the project sponsors, customers, and other stakeholders.

This is depicted as in Fig. 1.7.

The triple constraints of the core functions of scope, schedule and cost, and quality are shown in Fig. 1.8.

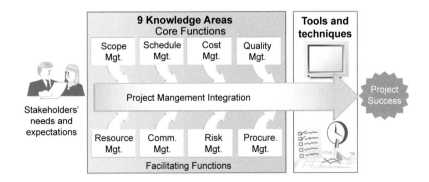

FIGURE 1.7

Project integration management.

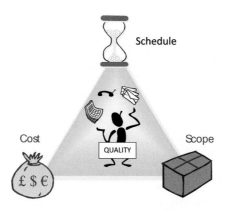

FIGURE 1.8

Triple constraint.

1.6.5 Operating cycles and turnaround phases

Definitions

Operating cycle[a]

This is the **run length** or time to the next turnaround and is determined from reviews of benchmarking, RCM, and RBI data (see Section 2.3).

Full cycle[b]

This is the time from start-up (online producing on-spec products) to start-up after the next turnaround. This is used to analyze the turnaround planned down days.

Turnaround/shutdown cycle[c]

This is the time from initiation of the next turnaround project to completion of the project (project duration).

The turnaround/shutdown cycle is divided into clear phases with each phase consisting of activities and targets for completion.

As in most projects, a project review or audit needs to take place at the end of each phase in the project, the most important being the readiness review/audit at the end of the detailed planning phase.

Front-end loading

This is a useful tool to assess performance in preparation for a turnaround (see Section 5.8.3).

Freeze date

This is the date when the scope is frozen, and the model **baseline** is set. This is done at least 6 weeks before the turnaround. Actual progress is then measured against this **baseline** plan.

Phases

Phase 1: Conceptual Development (Chapter 3)

Phase 2: Work Development (Chapter 4)

Phase 3: Detailed Planning (Chapter 5)

Phase 4: Pre-turnaround (Chapter 6)

Phase 5: Turnaround Execution (Chapter 7)

Phase 6: Post-turnaround (Chapter 8)

Fig. 1.9 depicts the above (including referenced superscripts).

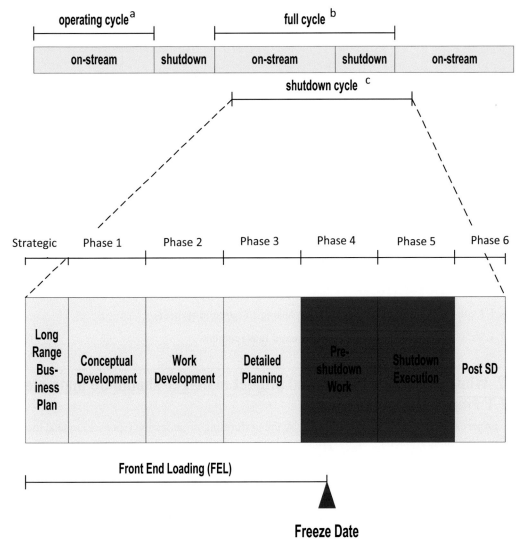

FIGURE 1.9

Cycles and phases.

Box 1.4: Turnaround phases compared

	Chevron		Singh[9]		Lenahan[10]	Capital Projects		Hey	
1	Long Range	1	Business Planning			Identify	Ends with Project Initiation Note (PIN)	Long Range	Ongoing for all turnarounds
2	Conceptual	2	Strategy & Alignment	1	Initiation	Assess	Ends with Initial Investment Decision (IID)	1	Conceptual: Ends with Initial Budget Decision (IBD)
3	Work Development					Select	Ends with Preliminary Investment Decision (PID)	2	Work Development: Ends with Preliminary Budget Decision (PBD)
4	Detailed Planning	3	Planning	2	Preparation	Define	Ends with Final Investment Decision (FID)	3	Detailed Planning: Ends with Final Budget Decision (FBD)
5	Pre-Shutdown Work	4	Pre-Turnaround			Execute		4	Pre-Turnaround Work
6	Shutdown Execution	5	Execution	3	Execution			5	Turnaround Execution
7	Post Shutdown	6	Post TA & Evaluation	4	Termination	Operate		6	Post Turnaround

---

Box 1.4 compares different approaches to phasing turnaround projects.

Box 1.5 shows the shell approach to turnaround management.

1.7 When and why a turnaround project manager should be appointed

1.7.1 Project manager definition

The project manager is the person assigned by the performing organization (sponsor) to lead the team that is responsible for achieving the project objectives.

1.7.2 Basic requirement

Ideally the project manager is appointed straight after the initial budget decision (IBD) gate. He should, however, be a key member of the development team before IBD.

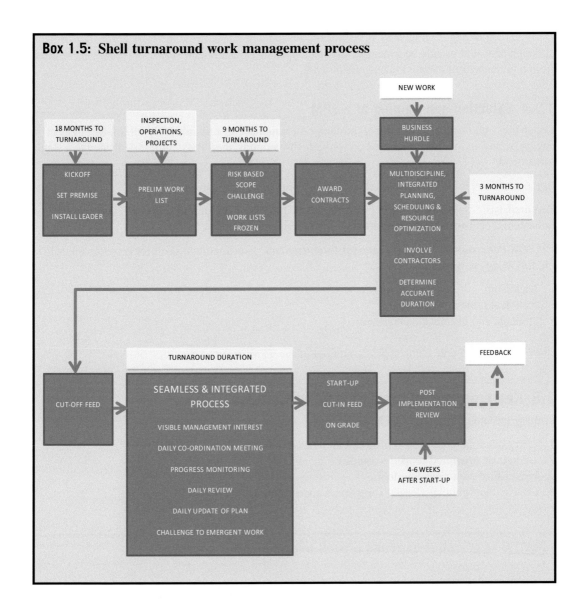

Box 1.5: Shell turnaround work management process

1.7.3 Alternative management approaches

The management of the turnaround and selection of those doing the work can vary considerably; Section 2.2 discusses this in more detail.

The TPM's title may vary according to the size of the turnaround and the contractual relationship with the sponsor. He could either be a member of staff or a contractor. On smaller turnarounds he could be referred to as the lead engineer or senior turnaround engineer.

If the amount of engineering work is significant, he could be drawn from the engineering department, although he would usually be a member of the maintenance department. However, he may be selected from a permanent turnaround management team.

1.7.4 Selection and training of a TPM

Depending on size, the TPM may be in charge of a number of turnarounds with a lead engineer in charge of each individual turnaround or, for large turnarounds, the TPM could be dedicated to a single turnaround.

In the first case, the TPM may also be referred to as a program manager.

Required basic interpersonal skills include leadership, team building, motivating, communicating, influencing, decision-making, negotiating, facilitating, coaching, etc.

PMBOK Paragraph 3.4 Project Manager Competencies[1] lists important skills for project managers in the following categories:

- leadership
- technical project management
- strategic and business management

Chevron describes core competencies required of the TPM in Box 1.6.

1.8 Key elements of a successful turnaround

Primary turnaround strategies include:

1. Carry out structured phased planning and execution
2. Undertake structured scope screening
3. Use critical path modeling tools
4. Choose the right experienced contractors and incentivize them

---

Box 1.6: Chevron turnaround project manager core competencies

- Stamina and adaptability
- Ability to see the big picture
- Able to manage group processes
- Able to test hypotheses on others
- Technical curiosity
- Concern for documentation and application of procedures
- Professional integrity
- Efficiency orientation
- Collaborative influence
- Team problem solving
- Technical skills and knowledge

Box 1.7: Key elements of success

   i. Establish a company specific set of processes, procedures, guidelines, and controls for turnarounds.
  ii. Implement strategic turnaround planning ensuring top management appreciation of the complexity and details involved, and focusing on well-defined goals and objectives.
 iii. Assign high-impact turnaround teams of dedicated and experienced staff and contractors with expert support.
  iv. Develop contracting strategies and plan to select the most qualified contractors and their management teams.
   v. Conduct turnaround planning preparedness reviews at the end of each phase of planning to ensure focused planning and efficient execution.
  vi. Ensure structured scope selection and control.
 vii. Utilize critical path modeling tools.
viii. Ensure a high level of productivity through well-defined roles and responsibilities.
  ix. Structure risk management systems to achieve a high level of safety and reduced risk in achieving agreed targets in time, cost, scope, and quality.
   x. Conduct post-turnaround evaluation based on benchmarking and key performance indicators to develop initiatives to eliminate weaknesses and improve future results.

Adapted from Singh R. World Class Turnaround Management. *Everest Press; 2000.*

5. Ensure staff are competent in both technical and management fields.

Box 1.7 lists a consultant's view of the elements of success.

1.9 Summary

Turnaround management is more than just managing the downtime for turnarounds. It is the art of maximizing profit over the life of the investment by optimizing uptime of the process at maximum throughput using strategic planning of turnarounds and structured implementation of each turnaround.

The chapters that follow will outline turnaround activities phase by phase.

References

1. A guide to Project Management Body Of Knowledge PMBOK (ANSI/PMI 99-001-2017).
2. Guidance on project management ISO 21500:2012.
3. Hey RB. *Turnaround, shutdown and outage management: effective planning and step-by-step execution of planned maintenance operations.* STO Asia 2014 IQPC.
4. Crowley K. *Big oil leaves analysts fuming about being in the dark on refinery outages.* Bloomberg; 2018.
5. Phillip Townsend Associates (PTAI). Available from: http://www.ptai.com/.
6. Solomon Associates. Available from: https://www.solomononline.com/.
7. Oirere S. *Africa: the challenge of investing in Africa's additional capacities hydrocarbon processing.* 2018.
8. Asset management ISO 55000:2014.
9. Singh R. *World class turnaround management.* Everest Press; 2000.
10. Lenahan T. *Turnaround management.* Butterworth-Heinemann; 1999.

Further reading

1. Hey RB. *The best turnaround ever.* Hydrocarbon Processing; 2001.
2. Hey RB. *Performance management for the oil, gas and process industries.* Elsevier; 2017.. In: *https://www.elsevier.com/books/performance-management-for-the-oil-gas-and-process-industries/hey/978-0-12-810446-0.*
3. Singh R. *Executive leadership essential to ensure world class turnarounds.* Global Turnaround and Maintenance Hydrocarbon Processing; March 2012.
4. Iyer KR. *Manage turnarounds effectively hydrocarbon processing.* 1996.
5. Banks A. *Geelong Refinery turnaround: precision planning and outcome.* 2017. Available from: www.vivaenergy.com.au/driven/innovation/geelong-refinery-turnaround-precision-planning-and-outcome.

Strategic planning

Strategic long-range business plan

2.0 Outline

This chapter discusses the development of a strategic plan for all turnarounds in a business unit or company. Strategic planning includes the development of both a long-term plan, typically 5 years, and a budget for all turnarounds in the business unit/company.

2.1 Basic requirements for optimization of the turnaround

2.1.1 Critical elements

A project management approach is essential for effective management of a turnaround (TA).

The following elements are especially critical:

1. Ensure clear alignment of TA statement of commitment with the corporate vision and mission (see Section 2.1.3)
2. Carry out structured phased planning and execution (see Section 1.6.5)
3. Undertake structured scope screening (see Section 3.2.1)
4. Apply front end loading (FEL) (see Section 5.8.3)
5. Use critical path modeling tools (see Section 5.3.1 and Appendix C)
6. Choose the best contractors with the most appropriate experience and incentivize them accordingly (see Section 3.9.4)
7. Ensure staff are competent in both technical and management fields (see Sections 3.6.2 and 3.6.3)
8. Use lessons learned for the next turnaround (see Sections 8.7.3 and 8.7.4)

All of these aspects should be well documented in a turnaround framework/manual.

This is summarized in Fig. 2.1.

2.1.2 Establishing the framework

Organizational process assets

PMBOK refers to organizational process assets as the plans, processes, policies, procedures, and knowledge bases specific to and used by the performing organization. Process assets include the

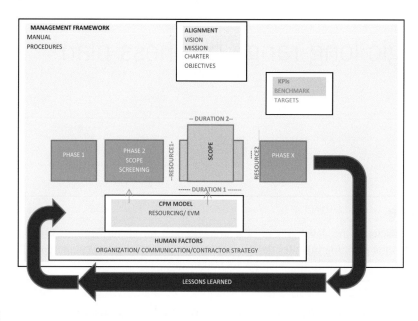

FIGURE 2.1

Turnaround management framework.[1]

organization's knowledge bases such as lessons learned and historical information. **Organizational process assets** may be grouped into two categories:

a. processes and procedures
b. corporate knowledge base

a. Processes and procedures

These include the framework, manuals, and procedures for managing turnarounds, including related maintenance and inspection procedures that are also relevant to online maintenance and inspection.

Grouping for turnaround project work could be:

I. Initiating and planning, which includes guidelines, standards, policies, procedures, and templates (risk register, WBS, etc.)
II. Executing, monitoring, and controlling, which includes change control, financial control, issues, and defect management procedures.

A list of turnaround management documents is given in Appendix E.

b. Corporate knowledge base/historical information

Historical information is grouped as follows:

I. Process records including catalyst performance, fouling, corrosion, coking, etc.
II. Inspection records, which include physical condition of all plant and equipment.

III. Maintenance records, which include past maintenance activities on all plants and equipment.
IV. Templates, which include previous turnaround critical path models, tools, and materials for specific tasks, isolation registers, etc.

2.1.3 Clear alignment of turnaround statement of commitment with the corporate vision and mission

Line of sight
Alignment between specific turnarounds and all turnarounds to the company mission and vision is required.

Alignment is as follows:

All turnarounds

1. Corporate vision and mission
2. Turnaround vision and mission

Specific upcoming turnaround

3. Project charter
4. Statement of commitment/turnaround philosophy
5. Perspective

All turnarounds

1. Turnaround vision
   Where we want to be.
2. Turnaround mission

   How we plan to get there.

This requires leadership from the top and buy in from key stakeholders in the business unit/company. Turnaround vision and mission needs to be in the introduction of the framework document.

A turnaround vision example is shown in Box 2.1.

A turnaround mission example is shown in Box 2.2.

Specific upcoming turnaround

3. Project charter and management plan
   This is to be signed off by the sponsor/decision maker (see Section 3.1.1 and Appendix A1).
4. Objectives (statement of commitment/turnaround philosophy)
   This is to be signed off by all key stakeholders: staff and contractors. Objectives include measurable project objectives and related success criteria: a commitment to achieve the scope, duration, and cost, with zero incidents, etc. (see Section 3.1.2).
5. Perspective
   This is the daily revenue/profit of the company that is lost for each down day (see Section 1.2).

Box 2.1: Turnaround vision example

TA Vision: Where we want to be

I. Improve the company's profitability by achieving **highest**:
   a. Plant **integrity** and
   b. Operational **reliability**
II. Become the **top** industry performer by achieving exceptional:
   a. **Safety and environmental** performance
   b. Reduced annualized turnaround **costs**
   c. Other efficiency improvements—expand
III. Accomplish these goals through:
   a. Cross-functional **teamwork** and effective **communication**
   b. Improved **productivity**
   c. **Innovation**

Box 2.2: Turnaround mission example

Mission: How we plan to get there

1. Provide **leadership and support** to turnaround team
2. Champion the use of **best** turnaround management **practices** and **innovation** to accomplish turnaround goals
3. Structure turnaround **work scope** development to minimize late add-ons or changes
4. Ensure **Health, Safety, and Environmental** values are incorporated in all turnaround phases
5. Establish measurable **key performance indicators**
6. Emphasize **timely and value-added planning, cost management, and controls**
7. Provide expeditious **decision-making**
8. Cultivate a **team work** environment with cross-functional teams and shared responsibilities
9. Empower turnaround teams reinforced with knowledgeable **tools and resources**
10. Implement a **contracting strategy** to select suitably qualified contractors
11. Embed **continuous improvement** through lessons learned, documentation, and information sharing

Adapted from Singh R. Achieve excellence and sustainability in your next turnaround. Hydrocarbon Processing; 2014[2].

2.2 Various structures for turnaround management
2.2.1 Management and work activity alternatives

The basic questions are:

1. Who will manage the turnaround?
2. Who will do the work?

The type and size of turnaround as well as the size of the company determines the approach to managing the turnaround and the contracting strategy for the turnaround.

Alternative management approaches are:

A. Manage the turnaround with only **company staff**.
B. Employ a **consultant,** independent of the contractor/s doing the work to manage the turnaround, but using staff members
C. Employ a **contractor** to manage and do all the work using its own staff and subcontractors.

Box 2.3 is an example of shutting down the whole complex and **managing the work using the company staff**. General contractors are employed for various areas and specialist contractors are employed by the company to support these area contractors. Full control is kept by company management, and, as a result, the majority of the risk stays within the company.

Alternatively, an alliance can be formed with a **consultant** to oversee contractors doing the work. In this case the client possesses the knowledge and experience of the plant that is to be shut down, whereas the partner has the turnaround project management and technical supervision experience. Box 2.4 gives an example.

Gas plants have the advantage of being able to independently shut down sets of trains and not the whole complex. An example of outsourcing the management to a **contractor** with experienced supervision staff and craftsmen is described in Box 2.5.

Work packages can vary from just one contractor doing all the work to several contractors doing work depending on area and/or specialist capabilities. The use of an agency to supply manpower is usually not recommended as productivity could be very low and the skills could be highly variable. Certain work such as electrical maintenance and inspection might be done with internal staff.

The Bahrain Petroleum Company organization breakdown structure shown in Box C4 gives a split categorized by type of equipment and specialist services. These are listed in Box 3.18.

Section 1.7 discusses the appointment of the project manager, irrespective of this being a staff member or consultant/contractor. Section 3.6.2 discusses the activities of the core team. The core team is appointed with or directly after the appointment of the turnaround project manager. Fig. 3.5 shows the relationships.

A policy team or steering committee would be the primary stakeholders. However, a single decision executive as well as a single turnaround project manager are required. The core team (sometimes referred to as the preparation team) would report to the turnaround project manager.

2.2.2 Transition from operations organizational structure to a project management structure

Should the turnaround require the entire refinery or complex to stop production, then it is necessary that those who normally operate and maintain the plant be part of the turnaround, either supervising turnaround activities or offering support roles such as those related to safety. Transfer of responsibility needs to be clearly identified.

Box 2.3: Chevron turnaround management approach

Cape Town Refinery SD2K turnaround

The plant
Chevron Cape Town Refinery is a 100,000 barrel per day integrated fuels refinery.

The approach to turnaround management
A complete refinery turnaround has the general manager as the turnaround project manager with departmental heads as process area managers. However, preparation is undertaken by a full-time core team that is established at the start of **Phase 1 Initiation: Concept development** as described in Chapter 3. The core team then supervises the contractors doing the work.

The selected turnover management team changes roles from routine operations management to a project team the moment the product stops flowing and reverts to operations mode the moment product starts flowing again. General contractors are employed for each area, while specialist contractors for refractory, scaffolding, carnage, etc. support each area contractor. The area contractors are under the direct day-to-day supervision of the company supervisors and management. Company supervision and inspection is supplemented from other Chevron refineries.

The following was the organizational breakdown structure for the Cape Town Refinery 2000 Turnaround (SD2K). Titles in boxed arrows were the line positions.

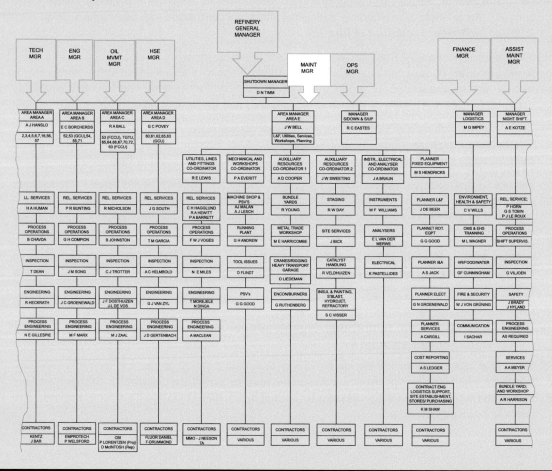

> **Box 2.4: South African Petroleum Refinery (SAPREF) turnaround management approach**[3]
>
> **SAPREF—Fluor promotion**
>
> *Turnaround management approach*
>
> An integrated shutdown was undertaken to implement scheduled inspections, operational cleaning, catalyst changing, and a number of de-bottlenecking projects. A fully integrated team of SAPREF and Fluor-SAPREF Alliance personnel carried out the work.
>
> *Client's challenge*
>
> SAPREF is a joint venture of Shell SA Refining and BP Southern Africa. The crude oil refinery is the largest in Southern Africa and represents 35% of South Africa's refining capacity. The refinery processes 24,000 tons of crude per day (200,000 barrels per day), producing 10 principal products in 46 grades.
>
> There are approximately 700 employees, 500 contractors, and 208 learners at the refinery.
>
> The project reinstated the mechanical and process integrity of equipment and piping systems in area E, for refinery Units 100, 200A, 200B, 300, 400, 500, and 700.
>
> The shutdown included tie-in of a new 24-meter vertical shell and tube heat exchanger, a "Texas Tower," in the platformer unit. The tower replaced 12 feed/effluent exchangers, increasing the reliability and capacity of the platformer. The tower was built from alloy steel by a South African manufacturer. It weighs 196 tons and has one of the longest rod-baffled bundles in the world.
>
> *Fluor's solution*
>
> A fully integrated team of SAPREF and Fluor-SAPREF Alliance personnel conducted the shutdown, which commenced with scoping, followed by a planning and preparation phase. It was critical that all parties review and approve the shutdown plan so that the responsibility for each activity was fully defined and the services to be provided by the team were well understood and integrated into a master schedule. The execution phase of the shutdown was carried out in three phases: mobilization, shutdown, and demobilization.
>
> The mechanical and process integrity of equipment and piping systems for seven units was reinstated. Mechanical, piping, rigging, and welding disciplines were involved, as were field engineering and quality control and assurance.
>
> Fluor's project control (planning) and warehousing groups set performance and professionalism benchmarks for the main mechanical contractors.
>
> *Conclusion*
>
> A 100% score on health, safety, and environment was achieved at each and every weekly client/management site inspection, which resulted in more than 300,000 work hours expended without a lost-time or medical recordable case.
>
> In addition, all phases of the shutdown were completed on schedule and below budget.

Time of handover

The time of handover of responsibility varies depending on the type of process.

Alternative handover times:

1. Handover occurs the moment the electricity stops flowing to the grid and handed back when the turbines are synchronized with the grid to produce electricity.

Box 2.5: Shell turnaround management approach[4]

Shell Pearl GTL Qatar

The plant

Pearl GTL is a fully integrated upstream-downstream development. It captures, in one operating business, the full gas value chain from offshore development through onshore gas processing, the conversion of gas to hydrocarbon liquids, and the refining to finished products. Up to 1.6 billion cubic feet per day of wellhead gas from 22 offshore wells is converted to gas-to-liquids using Shell's proprietary Shell Middle Distillate Synthesis process, built on 3500 patents. From this, a range of high-performing GTL products is created, from gas oil, kerosene, and base oil to naphtha and normal paraffins for the petrochemicals industry.

The Pearl GTL plant has 24 reactors, weighing 1200 tons a piece. They each contain 29,000 tubes full of Shell's cobalt synthesis catalyst, which speeds up the chemical reaction.

The approach to maintenance and turnaround management

After start-up of the new GTL complex in Qatar, Shell decided to outsource part of its maintenance and turnaround activities to contractors who have extensive experience in these fields.

This had the desired advantage of boosting the maintenance experience level in the complex instantly, and thus mitigated the risk of low levels of experience, as often happens in a new plant with staff going through a rapid learning curve.

The bidding process went through a prequalification stage to ensure the required capabilities were obtained and resulted in two long-term maintenance contracts being successfully established for the complex.

In such an instance, the contractor responsible for turnarounds ramps up its resources (both management and workers) for a particular turnaround, including project management expertise. The contractor is entirely responsible for the turnaround work to be carried out within an agreed time and budget with overall daily oversight by the client.

Power station example: Electricity Supply Commission (ESKOM)

A group of similar power stations, which are part of an electricity generating authority, have a dedicated turnaround team (supervisors and workers) under a dedicated turnaround project manager. This team goes from power station to power station on a continuous basis.

2. Handover to the project team occurs the moment the plant is cool enough to open up, and hand back occurs after boxing up of the plant.

Refining complex example: Bahrain Petroleum Company (Bapco)

*Bapco has a dedicated shutdowns and major maintenance department. The shutdown management team is responsible for a number of crude oil production trains as well as the power station and water treatment plant for the complex. A team of senior turnaround engineers under the departmental manager rotate to take charge of various turnarounds. One may be preparing for a turnaround while another is managing a current turnaround. The contractors have long-term contracts with the company and are allocated work activities by type of equipment (see **Box C4 Work Breakdown Structure Example**).*

2.2.3 Who will do the work?

Five different types of knowledge are required for turnaround management. These are:

1. Local—a knowledge of the history, layout, and peculiarities of the plant
2. Work management—a knowledge of tasks to be undertaken
3. Craft—a knowledge of details of work methods, tools, and equipment required, and time for completing the tasks
4. Specialist—a knowledge of specialist activities such as those relating to rotating machinery, fiscal metering, and licensed equipment
5. Management of discrete projects—project management knowledge and skills.

Key staff must know the history of the asset, and have been trained on technical aspects of maintenance/inspection and the project management of shutdowns.

Internal experience versus external experience in managing turnarounds needs to be evaluated for each particular turnaround.

Experience of the specific plant to be shut down requires internal staff who possess relevant knowledge of the particular plant. Specialist experience from contractors could be required for turbines, instruments, etc.

With respect to planning the turnaround, planners are required to have comprehensive experience of the type of plant to be shut down and be conversant with the relevant planning tools. Planners may be drawn from the operating company or could be primary contractors who will manage significant parts of the turnaround. Early involvement of the primary contractors would be of mutual benefit.

2.3 Strategic planning and run length determination

Traditionally, the run length was determined by statutory inspections of pressure containing equipment such as boilers, furnaces, pressure vessels, and heat exchangers, whereas downtime was reserved for completing those tasks that required the plant to be out of commission, as well as those items that maintenance had not been able to do with the plant on line. However, with the latest predictive tools, techniques and methods (risk-based inspection, reliability centered maintenance, etc.), the move has been toward simultaneously optimizing both the run length and downtime.

The following questions need to be raised before each turnaround:

1. What are the legal/insurance/industry standards requirements?
2. When are the required upstream shutdowns to take place?
3. What are the predicted market conditions? (in order to time the shutdown for minimum lost profit)
4. How are the units performing—mechanically and from a process point of view?
5. What other major shutdowns are planned in the country? (to avoid overloading available contractor resources)
6. What are the weather conditions at the time of the planned turnaround? (*For example, in Malaysia, turnarounds are not carried out during the monsoon season.*)
7. Should the whole plant be shut down, or just certain complexes or units?
8. What plant improvements/modifications are required during the shutdown?
9. What run length is being envisaged?

Factors that impact extension of turnaround run length periods are:

- Process severity issues
  - catalyst life
  - fouling
  - corrosion
  - coking
- Resistance to change by the leadership team and specialists.

Different approaches for determining the run length (operating cycle) are:

1. bench-marking
2. reliability centered maintenance
3. risk-based inspection

These are shown in Table 2.1.

Example

A Middle East oil refinery recently decided to have a run length of 6 years before their next turnaround. This is longer than the average, but is based on sound reliability and risk-based inspection data.

Table 2.1 Operation cycle determination.

	Benchmarking	**Reliability centered maintenance (RCM)**	**Risk-based inspection (RBI)**
Approach	Top-down generic	Bottom-up specific equipment	Bottom-up specific pressurized systems
Focus	Unit—hierarchical	Equipment—process flow	Pressurized systems
Comparison	Industry norms for type of unit	Statistical run time for equipment	Industry norms for pressure containing equipment
Application	Immediate	Long term	Long term
Data source	Last bench-marking study	Equipment vendor, industry and RD historical data	Plant historical data
Methodology	Solomon model (Profile II), EDC weighted	FMECA, etc.	Risk assessment software
Targets	Easily and quickly determined from industry norms	Determined after extensive data analysis	Set from output of risk assessment software
Objective	**Achieve industry norms for mechanical availability**	**Achieve optimum level of reliability for required process unit run length**	**Optimize period between internal inspections**

2.4 Establishing the long-term rolling turnaround schedule for all turnarounds in the business unit/company

Each planned turnaround in a business unit needs to be shown on a spreadsheet giving clear time indications graphically in bar chart form. (Microsoft Project is a popular tool for this.)

Box 2.6 shows an example of a simple 5-year look ahead of all turnarounds in the business unit.

Box 2.7 gives a detailed 2-year look ahead for the example shown in Box 2.6.

An example showing turnarounds for seven gas trains of a gas plant are shown in Box 2.8. This example gives actual and planned turnaround durations over 11 years as well as the unplanned down days. This is useful for performance assessment with respect to availability.

Box 2.6: Five-year look-ahead example

5 year refinery turnaround schedule		2018-2023	
	Major Turnarounds	*Cat Changes*	*Comments*
No 1 CDU	Feb-22		4 year interval for refinery area 1
No 1VDU	Feb-22		4 year interval for refinery area 1
No 1 FCCU	Feb-22		4 year interval for refinery area 1
No 2 CDU	Feb-20		4 year interval for refinery area 2
No 2 VDU	Feb-20		4 year interval for refinery area 2
No 2 FCCU	Feb-20		4 year interval for refinery area 2
Visbreaker	Feb-20		Furnace tube cleaning every 2 years depending on severity
Kero Hydrotreater	Feb-22	Feb 18/Feb 20/ Feb 22	KHT cat changes every 2 years
Diesel Hydrotreater	Feb-22	Feb 18/Feb 20/ Feb 22	DHT cat changes every 2 years
Naphtha Hydrotreater	Feb-22	Feb 18/Feb 21/ Feb 24	NHT cat changes every 3 years
Platformer (CCR)	Feb-22	Feb 18/	Cat changes - as required
ISOM Plant	Feb-22		4 year interval for complete refinery turnaround
LPG Merox Unit	Feb-22		4 year interval for complete refinery turnaround
Cat Poly Plant	Feb-22		4 year interval for complete refinery turnaround
Amine DEA	Feb-22		4 year interval for complete refinery turnaround
Amine MEA	Feb-20		4 year interval for complete refinery turnaround
No1 Sulphur Recovery Unit	Feb-22		4 year interval for complete refinery turnaround
No 2 Sulphur Recovery Unit	Feb-20		4 year interval for complete refinery turnaround
Tankage area 1	Feb-22		4 year interval for refinery area 1
Tankage area 2	Feb-20		4 year interval for refinery area 2
Flare Facilities 1	Feb-22		4 year interval for refinery area 1
Flare Facilities 2	Feb-20		4 year interval for refinery area 2
Steam Power Plant and Utilities 1	Feb-22		4 year interval for refinery area 1
Steam Power Plant and Utilities 2	Feb-20		4 year interval for refinery area 2
Cooling Water System 1	Feb-22		4 year interval for refinery area 1
Cooling Water System 2	Feb-20		4 year interval for refinery area 2

Box 2.7: Two-year look-ahead example

2 year refinery turnaround schedule	JAN	FEB	MAR	APR	MAY	JUN	JUL	AUG	SEP	OCT	NOV	DEC	JAN	FEB	MAR	APR	MAY	JUN	JUL	AUG	SEP	OCT	NOV	DEC	JAN	FEB
						2019												2020							2021	
No 1 CDU																										
No 1 VDU																										
No 1 FCCU																										
No 2 CDU												XX 28 days														
No 2 VDU												XX 28 days (benchmark target)														
No 2 FCCU												XXX 42 days (benchmark target)														
Visbreaker										X 7 days																
Kero Hydrotreater												X 27 days (benchmark target)														
Diesel Hydrotreater												X 30 days (benchmark target)														
Naphtha Hydrotreater																									31 days X	
Platformer																										
ISOM Plant																										
LPG Merox Unit																										
Cat Poly Plant																										
Amine DEA																										
Amine MEA												X 35 days														
No1 Sulphur Recovery Unit																										
No 2 Sulphur Recovery Unit												X 35 days (benchmark target)														
Tankage area 1																										
Tankage area 2												X 21 days														
Flare Facilities 1																										
Flare Facilities 2												X 14 days														
Steam Power Plant and Utilities 1																										
Steam Power Plant and Utilities 2												X 21 days														
Cooling Water System 1																										
Cooling Water System 2												X 21 days														

2.5 Turnaround critical success factors and key performance indicators

Critical success factors (CSFs) determine measurement areas on which to focus. Key performance indicators (KPIs) can then be determined more easily.

2.5.1 Critical success factors

Table 2.2 identifies the primary CSFs and related measurement areas.

2.5.2 Turnaround KPIs

Turnaround KPIs can be categorized as follows:

Categories

1. Predictability (variance)
   a. Measures a company's performance versus the company's estimates and targets set by the turnaround team
   b. Applies only to a specific turnaround
   c. Short term: target achievable with current tools/strategies

Box 2.8: Gas plant turnarounds: actuals and planned (2012)

Gas Plant planned and actual turnaround durations

BLACK: ACTUAL
RED: PLANNED AS PER WP&B

Unit		2008	2009	2010	2011	2012	2013	2014	2015	2016	2017	2018
T1	UNPLANNED						7	11	11	11	11	
3.3 MTPA	PLANNED				20.5/21	16/16	6	21				
C3MR	MONTH				MAY	MAY						
T2	UNPLANNED					7						
3.3 MTPA	PLANNED					21/21	6		21			
C3MR	MONTH					MAY			Mod Rep			
T3	UNPLANNED					7						
3.3 MTPA	PLANNED					16/16	18			21		
C3MR	MONTH					MAY				Mod Rep		
T4	UNPLANNED					7	11	11	11	11	11	11
7.8 MTPA	PLANNED					37/35	34		42			34
AP-X	MONTH		MAR			OCT			MOI			
T5	UNPLANNED					7	11	11	11	11	11	11
7.8 MTPA	PLANNED				9.1/23	40/37		42			34	
AP-X	MONTH		SEPT		MAY/NOV	OCT		MOI				
T6	UNPLANNED					8+	11	11	11	11	11	11
7.8 MTPA	PLANNED				13/35		34			42		
AP-X	MONTH			NOV	SEP/OCT		HGPI			MOI		
T7	UNPLANNED				14	24+	11	11	11	11	11	11
7.8 MTPA	PLANNED				14			34		42		
AP-X	MONTH				OCT			HGPI		MOI		

MOI — Major Overhaul & Inspection
HGPI — Hot Gas Path Inspection

Table 2.2 Critical success factors (CSFs) and measurement areas.

Critical success factor	Measurement area
Specific turnaround	
Duration	Oil/product out to on-spec product: days and annualized days
Turnaround cost	Actual and annualized
Predictability (variance)	Actual versus planned: man-hours, duration, and cost
Safety incidents	From oil out to on-spec: number and rate
Environmental incidents	From oil out to on-spec: number—total and major
Start-up incidents	Number and days lost
Additional work	Actual versus contingency
Leak-free start-up	Number of leaks causing delay in start-up and days lost
Productivity	Actual total man-hours on the job versus total clocked man-hours
Other	
Total maintenance cost	Turnaround and routine maintenance annualized
Frequency	Run length in months
Unscheduled shutdowns	Days lost during the run between turnarounds
Mechanical availability	Time available %—annualized
Initiatives	Time and cost savings—decontamination, modularization etc.
Front end loading	Preparedness for the next turnaround
Comparative statistics	Mechanical availability, turnaround duration (days), turnaround cost, run length (months), unplanned down days

2. Absolute (efficiency)
    a. Measures a company's effectiveness versus that of other companies (bench-marked)
    b. Applies to all turnarounds
    c. Long term: normally involves a paradigm shift by those involved for step improvement.

Next turnaround

1. Schedule
    a. Behind/ahead of critical path: hours
    b. Schedule variance (SV): actual versus planned duration
    c. Schedule performance index: alternative to SV
        i. Deviation from planned to date: earned value/planned value
2. Cost
    a. Cost variance (CV): actual versus planned costs
    b. Cost performance index: alternative to CV
        i. Deviation from budget to date: earned value/actual cost
    c. Deviation from budget: estimate at completion/budget at completion
    d. Additional work: actual versus contingency %

3. Scope
   a. Emergent work
      i. Man-hours as % of total turnaround man-hours: % (minimize)
      ii. Number of requests approved versus planned
4. Health, safety, environment (HSE) and quality
   a. Lost time incidents: Number (target: zero)
      i. Monitor lower level HSE indicators to prevent even one incident
   b. Flawless start-up (no incidents, no leaks): number (target: zero)
      i. Measure from "hand back" to "on-stream at required quality"
   c. Number of reworks (weld failures, etc.)
5. Productivity
   a. Productivity index
      i. Actual total man-hours on the job/total clocked man-hours.

Future turnarounds

1. Schedule efficiency: long-term turnaround duration target
   a. Turnaround duration relative to others with similar scope and complexity
2. Cost efficiency: long-term turnaround cost target
   a. Turnaround cost relative to others with similar scope and complexity
3. Plant availability (bench-marked)
   a. Maximum interval between turnarounds with
   b. Minimum unplanned down days and
   c. Minimum planned down days - annualized
4. Future run-time target
5. Safety, health, environment, and quality statistics
6. Front end loading
   a. Preparedness indicator

2.6 Turnaround bench-marking

The objective of turnaround bench-marking is to understand the effectiveness of the company's practice of planning, defining, and executing turnarounds.

The purpose of turnaround analysis is to:

- Measure turnaround target setting and performance of a company's turnaround management system relative to industry average
- Measure turnaround target setting and performance of an individual turnaround relative to the best turnarounds
- Quantify the level of definition or readiness to achieve "best in class" performance
- Identify gaps in targeted performance and gaps that may prevent "best in class" performance
- Develop a set of actionable recommendations for the teams to implement.

Box 2.9: Refinery turnaround duration and interval bench-marking example

	Refining Directorate			Solomon stats			Variance	
	Shed TA interval mths	Actual duration hours	Actual duration days	Annual-ized TA cost $/bbl	Actual TA interval mths	Days down for TA	Down time var % over	Interval var % under
Atmospheric Crude Distillation								
Unit 1	43	1488	62	11	49	28	225	88
Unit 2	36	998	42	10	48	32	128	75
Unit 3	36	976	41	10	48	32	126	75
Unit 4	43	936	39	11	49	28	142	88
Catalytic Cracking								
Units 51,52,53	34	2208	92	69	42	42	219	82

Comment

The above table compares actual turnaround intervals (operating cycle/run lengths) and durations for various process units to the benchmarked values published by Solomon Associates. "Downtime variance % over" depicts a worse duration than the industry norm and "Interval variance % under" depicts a worse run length than the industry norm.

Solomon[5] undertakes turnaround bench-marking specific to refining while Independent Project Analysis (IPA)[6] undertakes bench-marking on all refining and petrochemical industry turnarounds.

Box 2.9 gives an example of a refinery with respect to run length (interval) and duration of the turnaround.

Independent Project Analysis has a database of over 250 turnarounds.

The metrics provided to the client by IPA are shown in Box 2.10.

Box 2.10: IPA bench-marking reports[6]

Prospective Analysis: 4 to 6 months before start of turnaround	Readiness Review: 6 weeks before start of turnaround	Retrospective: Closeout after turnaround	System Bench-marking
Current Front End Loading (FEL)	Progress since earlier evaluation	Actual FEL	Metrics similar to Retrospective metrics except that the system is evaluated as well as consistency (or lack thereof) within the system.
Estimated FEL	Actual FEL and rating	Schedule performance	
Schedule performance risk	Updated schedule performance risk	Cost performance	
Cost performance risk	Updated cost performance risk	Total schedule predictability	
Estimated team integration	Updated team integration	Start-up schedule predictability	
Estimated use of Value Improvement Processes (VIPs)	Updated use of Value Improvement Processes (VIPs)	Cost predictability	
Recommendations	Updated recommendations	Team integration assessment	
		Safety performance Lost Time Incidents Recordable Incidents Environmental releases Calculated based on TA size	
		Work order growth	
		Lessons learned	

2.7 Summary

Basic requirements for a successful turnaround include:

1. Ensuring clear alignment of the turnaround statement of commitment with the corporate vision and mission
2. Implementing structured phased planning and execution
3. Undertaking structured scope screening
4. Applying FEL
5. Using critical path modeling tools
6. Choosing the right experienced contractors and incentivizing them
7. Ensuring staff are competent in both technical and management fields
8. Using lessons learned for the next turnaround.

The type of structure required for turnarounds within the organization are to be determined.

Run lengths of process units for the upcoming turnarounds need confirmation to determine likely dates for start of turnarounds in the next year. Factors in the decision-making process are:

1. market conditions/swings
2. unit performance—process and mechanical
3. legal requirements
4. capital planning
5. timing to minimize lost profit opportunities—value of downtime, cluster shutdowns, coordinate with other process plants

Tools for assisting in making decisions are:

1. unit process performance data
2. marketing strategy
3. product stock inventory
4. schedule of legally required inspections

Deliverables are:

1. annual shutdown schedule for all turnarounds in the next year
2. five-year rolling shutdown schedule
3. forecasted turnaround budget for next year for all turnarounds in that year

Timing is ongoing for strategic planning but comes to a head prior to approval of the following year's annual business plan and budget.

The next specific turnaround is determined and initiated, possibly using a project initiation note. This is the starting point for the next chapter.

References

1. Hey RB. *Turnarounds and profitability: optimizing maintenance costs STO Asia 2014 IQPC*.
2. Singh R. *Achieve excellence and sustainability in your next turnaround*. Hydrocarbon Processing; April 2014.
3. Fluor Website. Available from: www.fluor.com.
4. Shell Website www.shell.com.
5. Solomon Associates. Available from: https://www.solomononline.com/.
6. Independent Project Analysis (IPA). Available from: http://www.ipaglobal.com/.

Further reading

1. Hey RB. *Managing pacesetter plant shutdowns*. Project Management Institute South Africa (PMISA) 'Regional African Project Management'; 1999.
2. Oliver R. Complete planning for maintenance turnarounds will ensure success. *Oil & Gas Journal* April 2002.
3. McNair CJ, Liebfried KHJ. *Benchmarking—a tool for continuous improvement*. John Wiley; 1992.

Turnaround project planning

B

Phase 1 initiation: concept development

3.0 Outline

This chapter covers the key principles for turnaround planning, which are to:

- Establish a charter and philosophy as a basis for the upcoming turnaround
- Empower a cross-functional core team to plan and execute the turnaround
- Get early buy-in from top management with regard to schedule and cost plus other related issues
- Establish milestones for planning the turnaround
- Establish audits at key milestones to ensure that the project is on track.

These principles should be established for each turnaround in the concept development phase. This phase is generally short, typically 1–2 months in a petrochemical environment.

This phase ends when agreement has been reached on a budget proposal to enable the development of a plan and budget for a specific turnaround due to be undertaken in the following year, and for inclusion in that year's company business plan and budget (BP&B). This could be referred to as a **project initiation note** and includes all the necessary support documentation for a specific turnaround.

Phase 1 is discussed under the following headings as outlined in PMBOK.

1. Integration
2. Scope
3. Schedule
4. Cost
5. Quality
6. Resources
7. Communication
8. Risk (including health, safety, and environmental issues)
9. Procurement
10. Stakeholder management.

Table 3.1 specifically references PMBOK and ISO 21500 processes applicable to this phase.

Table 3.1 PMBOK and ISO 21500 processes.

	Phase 1: initiation—concept development	
No	**PMBOK**	**ISO 21500**
1	4.1 Develop project charter	4.3.2 Develop project charter
	4.2a Develop project management plan	4.3.3 Develop project plans
2	5.1a Plan scope management	4.3.11 Define scope
3	6.1a Plan schedule management	—
	6.2a Define activities	4.3.13 Define activities
4	7.1a Plan cost management	—
	7.2a Estimate costs (OOM)	4.3.25 Estimate costs
	7.3a Determine budget (BP&B)	4.3.26 Develop budget
5	8.1a Plan quality management	4.3.32 Plan quality
6	9.1a Plan resource management	4.3.17 Define project organization
	9.2a Estimate activity resources	
	9.3a Acquire resources	4.3.18 Establish project team
7	10.1a Plan communications management	4.3.38 Plan communications
8	11.1a Plan risk management	—
9	12.1a Plan procurement management	4.3.35 Plan procurements
10	13.1 Identify stakeholders	4.3.9 Identify stakeholders
	13.2 Plan stakeholder engagement	

3.1 Integration

3.1.0 Introduction and relevant project management (PM) processes

Two Integration documents are required to be produced in phase 1, namely the Project Charter and the Project Management Plan. In addition, a Turnaround Philosophy is also typically created for each Turnaround.

Standard processes are:

PMBOK	**ISO 21500**
4.1a Develop project charter	4.3.2 Develop project charter
4.2a Develop project management plan	4.3.3 Develop project plans

See Appendix A1 for details.

3.1.1 The project charter

The project charter is the document that formally recognizes the existence of a project. It establishes a partnership between the requesting and performing organizations. Chartering a project validates alignment between strategy and all ongoing work of the organization. It is formalized by the approval of an **initial budget decision (IBD)** and accompanied by all necessary attachments for each turnaround.

The charter should be agreed upon during the annual budget preparation stage and include the following:

1. Turnaround basis and summary
2. Company's turnaround vision and expectations
3. Goals and performance indicators
4. Work scope
5. Critical and risk areas
6. Organization and resources
7. Schedules and milestones
8. Budgets and cost estimates
9. Other
   a. Capital work (other budget/s)
   b. Communication and information management plan
   c. Training and orientation
   d. Health, safety, and environmental plan
   e. Materials management
   f. Contracting strategy and plan
   g. Quality control and inspection plan
   h. Operations plans and considerations
   i. Turnaround logistics plan.
10. Name and authority of sponsor (**decision executive**) and other person/s authorizing the project charter.

The approved long-range schedule, business plan, and charter are the initiating documents for phase 2.

3.1.2 Turnaround philosophy/statement of commitment

The plant owner (decision executive) lays the ground rules of the basic philosophy of the upcoming turnaround for agreement by management and all core team members.

The philosophy is then used by the core team to establish criteria for inclusion in the turnaround work list and to guide them throughout the turnaround.

A turnaround philosophy should:

- Minimize turnaround **duration** using critical path method (CPM)
- Meet turnaround **budget** using earned value method (EVM)
- Do only **those activities that require the plant to be shut down**
  - Complete all legally required inspections and repairs
  - Do what is required in the turnaround in anticipation of the agreed run length, with decisions being based on appropriate decision-making tools
- Be signed by interested parties.

A turnaround philosophy/statement of commitment should be formatted as follows:

Purpose

To achieve the targeted run length with minimum unplanned down days

Objectives

To achieve target:

- *Duration*
- *Cost*
- *Scope*
- *Lost time incidents*
- *Quality.*

Signed by interested parties.

An example is shown in Box 3.1.

3.1.3 Project management plan

The project management plan is a document or set of documents that defines how the project is to be undertaken, monitored, and controlled.

The turnaround project management plan includes:

1. Key events in preparing the turnaround, giving approximate time before the turnaround execution date
2. A series of milestones within phases.

Key events in preparing the turnaround, giving approximate time before the turnaround execution date are shown in Table 3.2.

A milestone schedule is then developed to ensure achievement of targets in preparing for the turnaround. This is discussed further in Section 3.3.

Box 3.2 shows a sequence of events for preparation of a typical turnaround for a particular refinery.

3.2 Scope
3.2.0 Introduction and relevant PM processes

Scope management in phase 1 entails setting the basic guidelines for scope inclusion in the turnaround and the process for collating and filtering work requests.

> **Box 3.1: Turnaround/shutdown philosophy example**
>
> **6 VDU T & I shutdown philosophy**
> Insure an incident-free and cost-effective shutdown by adhering to the following criteria:
> - Complete all justifiable work necessary to insure a safe, economical, and reliable 4-year plant run.
> - Complete all justifiable HAZOP, RED, and compliance items that require a shutdown for implementation.
> - Complete all justifiable capital work that requires a shutdown for implementation.
> - Minimize shutdown length by using critical path scheduling.
> - Establish and adhere to long-term schedules.
> - Utilize sound budgeting, planning, and control principles.
>
> By using the T&I Shutdown Manual as a guideline and the philosophy set forth above, we will establish the 1997, 6 VDU shutdown as a refinery pacesetter.
>
> Core team members:
>
> | P.G.Gay | J.J.Krett | A.R.Turki | D.P.Hughes | M.Ameen |
> |---|---|---|---|---|
> | Engineering | OPD | Mtce | TSD | RED |

The relevant standard processes are:

PMBOK	ISO 21500
5.1a Plan scope management	4.3.11 Define scope

See Appendix A2 for details.

3.2.1 Scope development 1

Poor work scope management has been recognized as the number one problem facing the turnaround management team.

Scope needs to be defined early on in the process and changes to scope after the scope freeze date needs to be strictly managed according to a scope change process.

All procedures related to scope management need to be reviewed and updated at this stage.

A narrative scope for the specific turnaround envisaged is to be included in the project charter. This is derived from a previous scope for a similar turnaround for this plant, obtained from the previous turnaround CPM model.

Table 3.2 Turnaround project management plan example.

Event size	Timing					
Small	12 months	6 months	4 months	3 months	2 months	2 weeks
Large	18 months	15 months	12 months	6 months	3 months	1 month
Phase	1: Conceptual	2: Work Dev	3: Detailed planning		4: Pre TA execution	
1. Integration	Appoint TAPM. Define TA charter/ philosophy/ goals.	Develop overall plan.		Integrate all turnaround activities.		Review checklists.
2. Scope: Work list preparation	Initial WL from insp/maint records. Set/amend scope challenge criteria.	Develop work packages.	Create WBS challenge and validate tasks.	Finalize work list. Final scope challenge.	Freeze work list.	Do required pre-turnaround work.
Projects/mods	Initial proj list from eng.		Integrate into plan.			
Major tasks	Identify major tasks from above.		Issue work packages to planning.			
3. Time: Planning	Milestone schedule. Initial plan template from previous TA model.	Define and categorize activities.	Detailed planning on CPM model.	Level resources and optimize model.		Ensure completion of all pre-turnaround work.
Scheduling	Duration from previous TA model.	Build turnaround model.		Finalize scheduling within agreed TA duration.	Finalize task method sheets. Freeze date and baseline.	
4. Cost: Costing	Cost from previous TA model escalated for budget approval.	Improve cost estimate. Prelim cost for approval.		Firm up budget. Final cost for approval.		Test cost management system.
5. Quality: HSEQ assurance	Review/update HSEQ standards and procedures.	Establish basic quality requirements. Review quality records.		Finalize updated QCPs.	Finalize QCPs.	Test QA system. QA in materials received at site.

Table 3.2 Turnaround project management plan example.—cont'd

Event size	Timing					
Small	**12 months**	**6 months**	**4 months**	**3 months**	**2 months**	**2 weeks**
Large	**18 months**	**15 months**	**12 months**	**6 months**	**3 months**	**1 month**
Phase	**1: Conceptual**	**2: Work Dev**	**3: Detailed planning**		**4: Pre TA execution**	
6. Resources/ Organization	Establish core team. Trade and special service resource planning.	Establish org breakdown structure Do craft resourcing estimate.	Equipment and tooling requirements.	Finalize resource requirements.	Mobilize execution team.	Team building and training. Contractor mobilization.
7. Communication	Establish communication structures and lines of communication.	Develop logistics plan.		Finalize updated procedures. Publish plot plan.	Initiate flawless turnaround awareness program.	Final execution plan published. Briefings and notifications.
8. Risk	Draft risk register from previous TA.	Identify/analyze/ evaluate/ respond/treat/ document risks.	Assess and mitigate risks.	Work list contingency review. Update risk register.	Final risk register.	HSE control enforced. Change control implemented.
9. Procurement	Review long lead items and contracting strategy.	Constructability alternatives. Identify major contractors and put out to tender. Order long-lead items.	Award major contracts. Identify specialist contractors and put out to tender.	Award to specialist contracts.		Mobilize contractors. Prefabrication completed.
10. Stakeholders	Identify stakeholders. Obtain decision exec approval for project charter/ budget.	Other stakeholder buy-in.		Obtain decision exec approval for final project charter/budget.	Decision to go ahead by decision exec.	Other stakeholders notified of go ahead.
Audits	1	2		3: FEL	4: TRR	

Box 3.2: Turnaround preparation sequence sample: planner's activities

RS: Reliability Services Section; EPRRC: Equipment Purchase and Repair Recommendation Committee

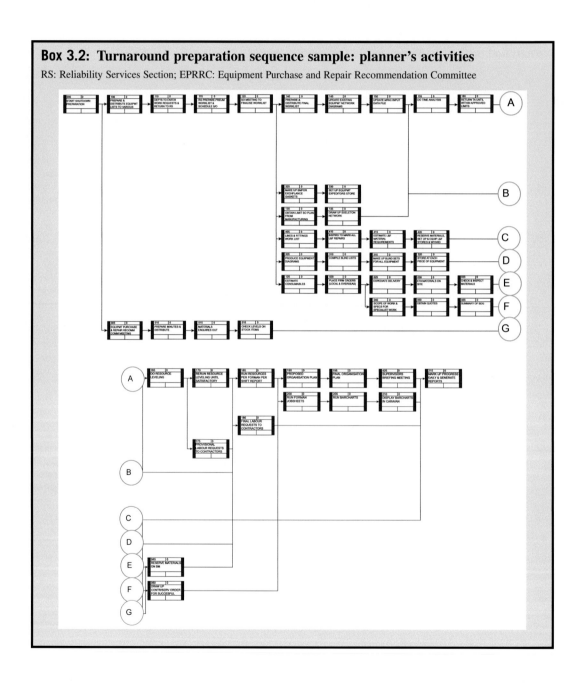

Contacts in each department need to be established. They are responsible for giving leadership and timely development of the turnaround work scope.

The work scope review team needs to be established at this time (see Section 3.6.5). Their role is to draft/amend the policy and guidelines/methodology for turnaround scope inclusion.

These guidelines should include the following in order of prioritization:

Priority 1: legal and insurance requirements

1. Compliance with regulatory agencies and other certification requirements—legal, insurance, etc.
2. Safety, health, environment, or quality items requiring a shutdown.
3. Items that cannot be deferred without an unacceptable level of risk.

Priority 2: corporate policy requirements

4. Consistent with corporate direction statement, turnaround philosophy/statement of commitment, and agreed budget goals derived from the business plan.

Priority 3: reliable run length—100% operating factor with no unplanned shutdowns

5. Inspection requirements to maintain an acceptable level of reliability.
6. Work that cannot be done during routine operations, thus requiring a shutdown for the work to be completed.
7. Work that has a potentially higher safety and operational **risk** if performed during routine operations.
8. Common utilities work such as electrical substations, sewages, etc.
9. Electrical and instrumentation upgrades and modifications.
10. Mechanical integrity work items to ensure operational reliability between scheduled turnarounds.
11. Opportunistic work—more efficient and cost-effective execution during a turnaround.

Priority 4: increased profit opportunities

12. Energy efficient items that could improve plant economics.
13. Process operating changes justified and approved.
14. Capital work tie-ins and modifications with existing facilities.
15. Long-term programs for abatement, corrosion, civil work, etc.

Inputs into the development of the work list include:

- Turnaround files and reports (including lessons learned from previous turnarounds)
- Maintenance work awaiting the turnaround

- Inspection records
- Process records (catalyst effectiveness, fouling, plant upsets, etc.)
- Best practices
- Process safety management inputs
- Core team member inputs
- Environmental and safety member inputs
- Reliability statistics.

Fig. 3.1 demonstrates a feasible decision process.

Potential work scope problems are listed in Box 3.3.

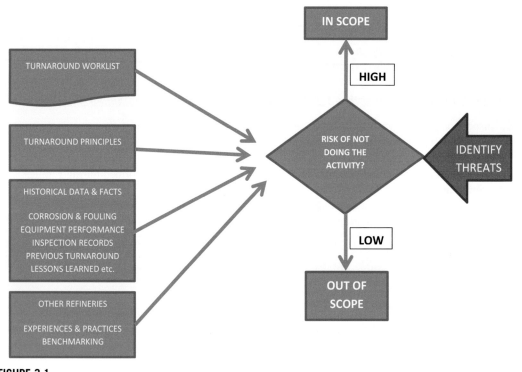

FIGURE 3.1

Scope decision process.

Box 3.3: Work scope problems
- Lack of formal work scope development procedures and guidelines
- Absence of agreed-upon work scope development criteria
- Lack of work scope prioritization
- Work scope requests are too generalized and lack detail
- Late and incomplete work scope
- Constantly changing work scope
- No contingency planning for potential work scope growth
- Reduced work scope to meet target budgets and then adding them back again as soon as the turnaround starts
- Lack of equipment history to develop a good work scope
- Slow decision-making to approve changes and inspection recommendations.

3.2.2 Turnaround work order listing process

Work orders for actioning during a turnaround could be accumulated in a computerized maintenance management system (CMMS). A sample process is shown in Box 3.4.

Box 3.5 gives an example of the turnaround planning process once an item has been approved for inclusion in the turnaround.

Alternatively, a spreadsheet listing could be used. An example of part of a listing is shown in Box 3.6. Note that the final signed off list is the frozen scope for entering in the CPM model.

3.3 Schedule
3.3.0 Introduction and relevant PM processes

In phase 1, key activities have to be defined and include times for completion.

Relevant standard processes are:

PMBOK	ISO 21500
6.1a Plan schedule management	–
6.2a Define activities	–

See Appendix A3 for details.

3.3.1 Milestone schedule

One of the first items for the core team to develop is a milestone schedule for the particular turnaround. CPM models and milestones from the last turnaround are good starting points for this.

Box 3.4: Turnaround/shutdown planning acceptance process example (SAP/Primavera)

PCR: Project Change Request

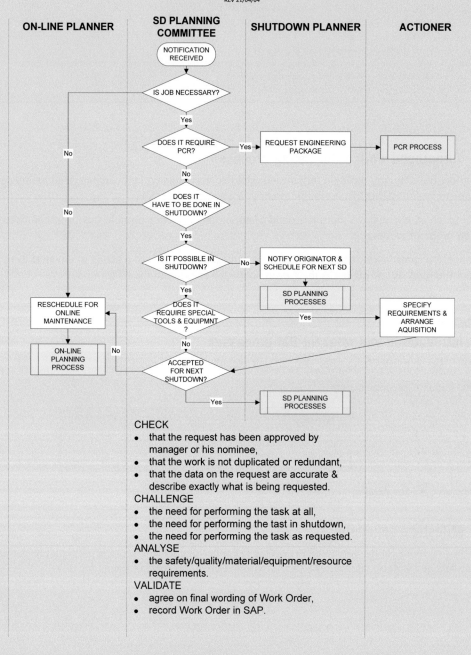

WORK FLOW
SHUTDOWN PLANNING ACCEPTANCE PROCESS
REV 21/04/04

ON-LINE PLANNER	SD PLANNING COMMITTEE	SHUTDOWN PLANNER	ACTIONER

NOTIFICATION RECEIVED

IS JOB NECESSARY? — No

DOES IT REQUIRE PCR? — Yes → REQUEST ENGINEERING PACKAGE → PCR PROCESS

No

DOES IT HAVE TO BE DONE IN SHUTDOWN? — No

Yes

IS IT POSSIBLE IN SHUTDOWN? — No → NOTIFY ORIGINATOR & SCHEDULE FOR NEXT SD → SD PLANNING PROCESSES

Yes

RESCHEDULE FOR ONLINE MAINTENANCE

DOES IT REQUIRE SPECIAL TOOLS & EQUIPMNT? — Yes → SPECIFY REQUIREMENTS & ARRANGE AQUISITION

No

ON-LINE PLANNING PROCESS

ACCEPTED FOR NEXT SHUTDOWN? — Yes → SD PLANNING PROCESSES

CHECK
- that the request has been approved by manager or his nominee,
- that the work is not duplicated or redundant,
- that the data on the request are accurate & describe exactly what is being requested.

CHALLENGE
- the need for performing the task at all,
- the need for performing the tast in shutdown,
- the need for performing the task as requested.

ANALYSE
- the safety/quality/material/equipment/resource requirements.

VALIDATE
- agree on final wording of Work Order,
- record Work Order in SAP.

Box 3.5: **Turnaround/shutdown planning process example (SAP/Primavera)**

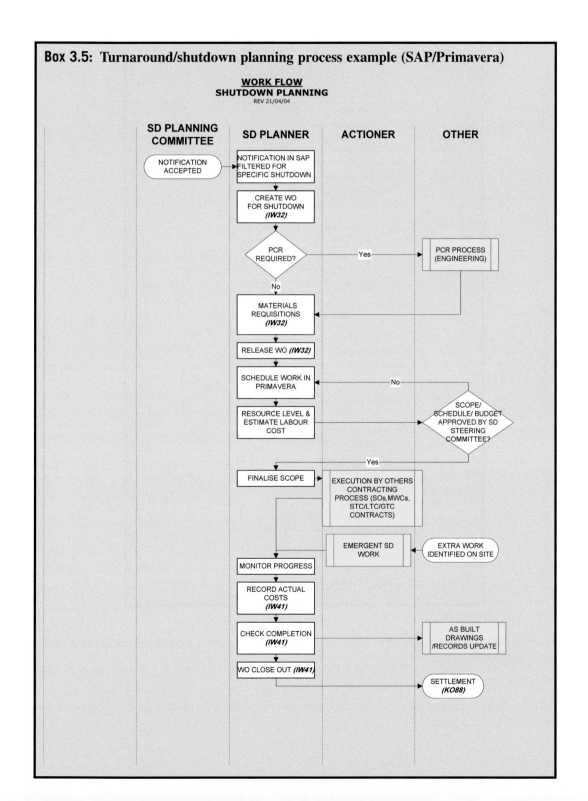

Box 3.6: Work list example

		(Worklist for sign off before new revision is issued)(B Chavda)		
SOURCE	STEP	WORK REQUIREMENTS	REV	QCP
Plant	5	Visbreaker Unit		
Plant Owner:		Bharat Chavda		
PLANT5		PLANT 5 GENERAL		
RSINSTR	1	Check & service all ESD system loops.		
RSINSTR	2	Anticipate broken/faulty pressure gauges and replace		
5C1		VISBREAKER FRACTIONATOR		
INSP	1	Open column and sandblast unlined section		
INSP	3	Remove all sections of trays 11 to 14 and the chimney to workshop for inspection and repair of corroded or non-standard components.		
INSP	4	Remove temporary liner plates and nozzle trim plates from nozzles T1, T3, and T5 (2" and 4" nozzle pairs at N. and N.w. orientations respectively and renew in accordance with Q.C.P. 11345-96)		
INSP	5	Monitor thickness of distributor pipe and pipe support above tray 14.		
INSP	7	Anticipate possible cracking of sieve trays 7 to 10 due to further in service embrittlement of 12% chromium steel material		
INSP	8	Renew nut welded on top of bottom manway cover. (QCP 11345-96)		
INSP	10	Provide staging for access to all Ramshorn Nozzles at lower north side and clean away all leaking product (hydrojet) for inspection of nozzles		
MOC 51754	1	Redesign of chimney tray inside 5C-1	Del Rev6te	
MOC 51754	2	Replace the chimney tray and change the material from 12Cr to S/S.		
MOC 51867	1	Replace the top section of column		
MOC 51867	2	Renew top head and shell above tray 21 or upgrade with new TP405 cladded carbon steel.	Del Rev6te	
OPS	1	Trim sharp point on tray 10 which hinders access into column		
OPS	2	Fully weld chimney tray (except manway)		
RSINSTR	2	Overhaul level switch 05LSH-015 (Jo-bell)	Del Rev6te	
RSINSTR	3	Overhaul level switch 05LSL-016 (Jo-bell)	Del Rev6te	
RSINSTR	4	Remove/replace for inspection 05TW-008		
RSINSTR	5	Remove/replace for inspection 05TW-108		
RSINSTR	6	Remove/replace for inspection 05TW-114		
RSINSTR	7	Remove/replace for inspection 05TW-130		
5D1X		FRACTIONATOR ACCUMULATOR		
INSP	2	Carry out TOFT on area of shell plate to test for hydrogen blistering		
RSINSTR	1	Overhaul & calibrate level-trol 05LT-003.	Del Rev6te	
RSINSTR	3	Overhaul level switch 05LSH-017 (Jo-bell)	Del Rev6te	

Rev. 6temp 12 October 1999   Report Date 13/10/1999 16:20:04   Page 48 of 97

The milestone schedule must have clear demarcation as to what is required in each phase of the turnaround.

A typical milestone schedule is shown in Box 3.7.

An engineering work order log is suggested for work orders that are expected for the turnaround (see Box 5.3 for an example).

Box 3.7: Milestone schedule example

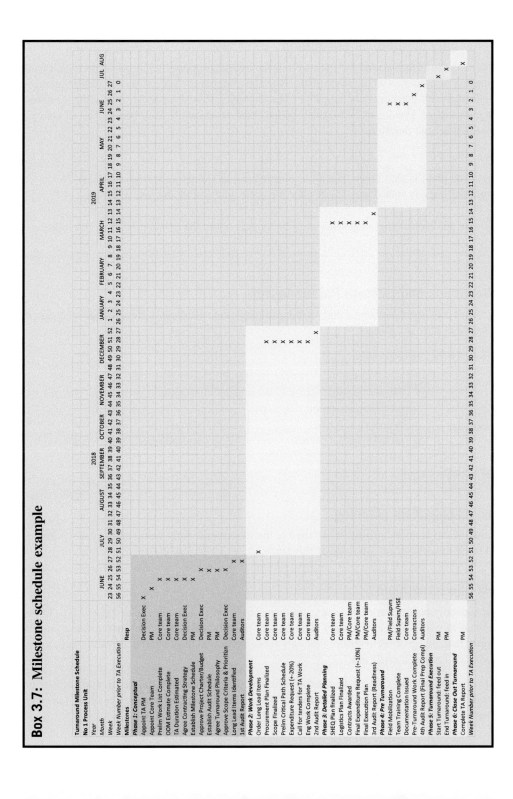

3.3.2 Preliminary duration determination

The duration of the next turnaround for inclusion in the budget proposal is initially determined from past experience of turnarounds on the same plant, as well as comparative durations for similar plants in the industry from bench-marking data (see Box 2.9). Strategies to achieve better than industry norms are decided at this stage. Previous CPM models of the same plant should have been saved to assist in determining the potential duration.

Two contradictory factors are:

1. The longer the plant is shut down, the more improvements and upgrades can be made
2. The longer the plant is shut down, the greater the costs and greater the lost revenue.

A primary factor in determining what work is to be carried out is the existing level of reliability.

3.4 Cost

3.4.0 Introduction and relevant PM processes

In phase 1, cost management is planned and an order of magnitude estimate is prepared for inclusion in the annual business plan and budget.

The relevant standard processes are:

PMBOK	ISO 21500
7.1a Plan cost management	—
7.2a Estimate costs (OOM)	4.3.25 Estimate costs
7.3a Determine budget (BP&B)	4.3.26 Develop budget

See Appendix A4 for details.

3.4.1 Cost management planning

The cost management planning process is detailed in the turnaround framework, and any amendments to this need to be identified and approved at this stage.

3.4.2 Cost of a turnaround

Turnarounds typically occur every few years and thus should be classified as nonroutine items in the company annual business plan and budget. The budget must be related to the agreed scope after a rigorous scope challenge process to ensure that only those items that have to be done in the turnaround are fully budgeted for and carried out.

One must keep in mind the **cost of lost production** as a result of a turnaround **and** unplanned downtime relative to the **cost of the turnaround**. If there is too much cost cutting for the turnaround, higher unplanned downtime may occur (see Sections 1.2 and 1.3).

3.4.3 Cost estimates and budget

Generally there are three estimates as the scope of the project is developed.

1. Initial Estimate: an order of magnitude (OOM) estimate ($\pm 30\%$) for inclusion in next year's budget
2. A preliminary estimate if the OOM estimate is inadequate ($\pm 20\%$)
3. A final estimate based on the CPM model ($\pm 10\%$).

The initial cost estimate for inclusion in the budget proposal is based on estimates of previous turnarounds on this plant escalated to the turnaround date. The IBD is the outcome.

Cost estimating is ideally carried out by a cost engineer with local experience.

Escalation
Previous estimates need to be escalated to the turnaround date as per industry norms.

Order of magnitude estimate process

1. Copy the actual turnaround model from the previous turnaround.
2. Estimate manpower (man-hours) required for items not covered in the previous turnaround but that are included in the following turnaround, and remove items not required.
3. Calculate total man-hours required and multiply by a productivity factor (normally between 1.4 and 2). See Sections 5.6.5 and 7.6.2 for details on productivity.
4. Escalate man-hour costs to date of turnaround and calculate average for all trades.
5. Calculate total manpower cost (total man-hours $\times$ average cost per hour).
6. Multiply total manpower cost by a factor to include materials and overheads (determined by a cost engineer with local knowledge).

Box 3.8 gives an example of being pressured to reduce a turnaround budget prior to submission of the company annual budget.

3.5 Quality
3.5.0 Introduction and relevant PM processes

The planning of quality management for the specific turnaround is also carried out in phase 1. This is based on company and other quality and inspection standards.

Box 3.8: Pressure to reduce turnaround budget

The initial budget for a major refinery turnaround was submitted to the board of directors for approval as part of the annual company business plan and budget (BP&B). The board chose to arbitrarily cut the turnaround budget in half.

The turnaround proceeded in the budget year, but the actual cost turned out to be double the approved budget.

Comment

The submission of a formal project charter with relevant attachments would have clarified all requirements for approval by the decision makers, which could have reduced the risk of a discretionary cut in the turnaround budget.

The related project standard processes are:

PMBOK	ISO 21500
8.1a Plan quality management	4.3.33 Plan quality

See Appendix A5 for details.

3.5.1 Quality management planning

Standard quality procedures in the turnaround framework are used as a basis for the quality management approach to be adopted for the coming turnaround.

Quality procedures and specific quality control plans (QCPs) are amended as required.

Certification of coded welders commences and the coded welder register is updated (as required by various standards such as ASME).

A register of qualified tradesmen could be established. Box 3.9 describes problems related to sourcing qualified tradespeople.

The quality of materials delivered for the turnaround could create a headache even when all efforts have been made to get the materials to the site in time. An example of such a material issue is described in Box 3.10.

3.5.2 ISO 9000: 2015

Company or business unit certification to ISO 9000 requires that all systems and processes be recorded and applied, and a continuous improvement process be in place. Fundamentals include the application of lessons learned from previous turnarounds. Appendix E discusses this standard in more detail.

3.5.3 Materials quality management

The management of the quality of materials for most large process plants is carried out within the materials module of the company enterprise resource management (ERM) system. The required

---

Box 3.9: Bahrain Petroleum Company's turnaround resourcing problem

In the 1990s, the Bahrain Petroleum Company (Bapco) was experiencing problems with sourcing qualified fitters, welders, and other trades from India for turnarounds.

A rumor began circulating that a carpet fitter had arrived on site as a so-called "qualified fitter" for a refinery turnaround.

Bapco subsequently decided to trade test all craftsmen prior to allowing them to work on turnarounds. All craftsmen were registered in a company register and their qualifications limited to a few years, after which they were retested. This had always been a requirement for welders, but it was then expanded to all trades working on a turnaround.

In addition, each approved tradesman had to carry an identity card showing his qualifications with the relevant expiry date.

Comment

Petronas's requirements to enter its offshore facilities include a "safety passport," as well as all qualifications required to perform the relevant activity on site (permit to work certificate, etc.).

> **Box 3.10: Bahrain Petroleum Company (Bapco) materials quality problem**
>
> In preparation for a major refinery turnaround, Bapco identified the requirement for replacement piping for its main cooling water intake lines from the sea.
>
> The piping was to be cement-lined carbon steel (as per the company specifications) and was imported from the United Kingdom (UK). A third-party inspection and expediting contractor was appointed in the UK to ensure the piping met specifications and was shipped timeously for the start of the turnaround. The piping duly arrived 2 weeks before the turnaround as planned, but on inspection of the lining, hairline cracks were found throughout most of the piping, which could have been a major setback. However, the inspection department was consulted as to alternatives available and it was decided that all the cement lining be removed and a polypropylene lining applied, using a special sprayer that was available in the region.
>
> In spite of what could have been a huge delay, the turnaround started on schedule and the polypropylene-lined pipes were used to replace the old corroded cement-lined steel pipes.
>
> **Outcome**
> The third-party inspection and expediting company in the UK was no longer used, and the salt water piping specification was changed to include lining with polypropylene instead of cement.
>
> **Comment**
> One needs to weigh up the cost of company's own manpower to be sent for FAT overseas versus cost of third-party inspectors.

standards for purchase of materials and spares for oil and gas plants are embedded in this module. Box 3.15 gives an example.

3.6 Resources

3.6.0 Introduction and relevant PM processes

In phase 1, the project manager and core team are appointed. Early booking of specialist services and equipment may also be required.

Relevant standard processes are:

PMBOK	ISO 21500
9.1a Plan resources	–
9.2a Estimate activity resources	
9.3a Acquire resources	4.3.15 Establish project team

See Appendix A6 for details.

3.6.1 Project management appointments

Senior management first forms a **steering committee** (policy team), which typically consists of senior management of the facility. The steering committee provides direction and guidance to the turnaround project manager and core team to ensure that the turnaround meets the needs of the business. The **decision executive** is typically the chairman of this committee. It must also ensure that the scope and budget for the turnaround are in alignment (see Box 3.8 for an example of misalignment.).

Senior management appoints a **turnaround project manager** and forms a **core team** specifically for the upcoming turnaround. Picking the best people for the core team will maximize the likelihood of a successful turnaround.

Selected staff would include those with the following:

1. A thorough knowledge of process operations within the plant that is to be shut down
2. A thorough knowledge of the inspection requirements for the relevant plant
3. A thorough knowledge of maintenance of the relevant plant
4. Experience of shutting down and starting up the relevant plant
5. Expertise in modeling previous turnarounds of the particular plant.

3.6.2 Human resource planning

Turnarounds require input and support from all departments affected by the turnaround. Thus a core team, which has been appointed for a particular turnaround and led by the turnaround project manager, must represent these key departments. This team must be able to lead the planning and execution of the turnaround and be empowered by the decision executive to get on with the job. Continuous communication between the core team and the decision executive is required to ensure concurrence, especially before moving on to the next phase. This team should be formed at least a year before the start of the turnaround execution.

A likely core team will consist of:

- Turnaround project manager
- Planning engineer
- Operations representative
- Design engineer
- Inspector
- Maintenance engineer
- Materials specialist
- Process engineer
- Health, safety, and environment coordinator
- Logistics coordinator

3.6.3 Core team job functions

Detailed job functions are described as follows:

Turnaround project manager (TPM)—team leader

1. Get process planning input on suitable turnaround dates and the cost of downtime
2. Develop a milestone schedule for the particular turnaround with input from the core team and others
3. Initiate the order of long lead items required for the turnaround
4. Maintain the updated turnaround work list

5. Schedule and chair all turnaround meetings
6. Communicate with all necessary stakeholders to achieve the turnaround objectives
7. Develop manpower requirements for the turnaround
8. Prepare contracts and recommend awards for turnaround work
9. Write expenditure request and submit for approval
10. Conduct job walk-throughs with execution phase supervisors prior to start of turnaround
11. Manage the turnaround during execution
12. Collate the final turnaround report.

Planning engineer

1. Act as TPM's right-hand man and stand in for him whenever he is unavailable
2. Record minutes of all turnaround meetings
3. Input the milestone and other schedules into the planning model
4. Develop the planning model using CPM and EVM
5. Produce cost estimates
6. Produce critical path diagrams
7. Produce resource histograms
8. Do "what ifs"
9. Dump data to and from the planning model
10. Monitor materials purchasing and delivery progress
11. Monitor turnaround progress and expenditure
12. Produce tailor-made reports
13. Advise core team and decision makers on the optimum use of the planning and costing tools.

Operations representative

1. Provide input on operational difficulties and safety issues
2. Provide blind lists, valve repair lists, and flange repair lists
3. Provide input on identifying pre-turnaround work
4. Provide input on shutdown and start-up sequencing
5. Coordinate operations support for plant preparation for turnaround
6. Train operations personnel for turnaround duties, such as fire watch, hole watch, etc.
7. Coordinate operations activities during start-up
8. Participate in Hazops, MOC reviews, etc.

Inspector

1. Provide a comprehensive list of inspection recommendations including compliance items
2. Represent all inspection groups, including rotating machinery, pressure equipment, electrical, and instrument inspections
3. Coordinate on-stream inspection with turnaround planning
4. Participate in evaluating and resolving inspection items
5. Coordinate inspection efforts during the turnaround in providing inspection support at critical steps so as not to interfere with progress.

Design engineer

1. Execute duties of turnaround project manager if also selected for this post
2. Provide engineering contacts for all capital jobs
3. Update the engineering work request log for turnaround work
4. Manage MOCs, Hazops, etc.
5. Provide planning input for capital projects.

Maintenance engineer

1. Execute duties of turnaround project manager if also selected for this post
2. Provide data and experience from previous turnarounds and online maintenance
3. Provide input on nonturnaround and pre-turnaround work
4. Participate in Hazops, MOC reviews, etc.

Materials specialist

1. Coordinate all turnaround materials purchasing activities (including liaising with a skilled focal person from the company procurement team who has extensive experience in dealing with vendors)
2. Arrange expediting where necessary
3. Log and monitor all ordering progress
4. Notify turnaround project manager timeously of delivery delays so that fast tracking can be initiated.

Process engineer

1. Provide early input for all process-related work, including process performance, environmental legislation compliance, catalyst regeneration and/or replacement, hazardous waste treatment and disposal, capital projects, Hazops, MOC requests, etc.
2. Input into procedures for turnaround, start-up, plant preparation, hazardous waste disposal, etc.
3. Participate in critical activities requiring process engineering input during the turnaround
4. Assist with start-up of process.

Health, safety, and environmental (HSE) coordinator

1. Coordinate all HSE issues
2. Advise on solutions to HSE issues (including streamlining permits to work and assess processes).

Logistics coordinator (initially part time and later full time)

1. Drawing up site master plot plan (see Section 5.7.3)
2. Undertake other activities covered in Section 4.7.1.
3. Work closely with materials specialist, especially related to delivery of major items of equipment (see Section 5.9.5).

As previously stated, the core team must have a functional leader or facilitator, normally referred to as the turnaround project manager (TPM). This person schedules meetings, retains master work lists and schedules, and develops updates for the decision executive. This job could be allocated to an engineer from engineering or maintenance who has a thorough knowledge of the plant to be shut down.

Team training or "building" may be required to get all members working well together.

3.6.4 Work scope review team responsibilities

The work scope review team needs to be multidisciplinary and selected from the steering committee, core team, and any other interested party. Fig. 3.5 shows relationships.

Work scope review team responsibilities include:

- Establish TA specific work scope criteria (see Section 3.2.1)
- Assign work scope development responsibilities
- Decide work scope cutoff dates in consultation with responsible groups
- Ensure early work scope finalization and planning considerations for critical, complex, or risky work items such as heaters, reactors, compressors, etc.
- Prioritize work scope in order to meet budgetary or schedule mandates
- Approve or disapprove any late work scope additions or changes.

3.6.5 Key attributes of a scope review team and core team member

Members of these teams need to be carefully assessed for maximum benefit to the turnaround project.

Some of the key attributes of a top-rated team member are:

- Ability to see the big picture—able to see how jobs impact each other, explain the relationships between jobs, home in on the root cause of problems, and activate a solution
- Problem solving tenacity—works on a problem as part of a team until it is solved, regardless of barriers and other restraints
- Stamina and adaptability—remains calm and patient in situations of continuing high stress while maintaining a high level of performance
- Communication skills—promotes teamwork and commitment to shared goals, cooperation between participants and keeps all relevant stakeholders in the picture
- Conflict management—able to resolve conflict, preferably in a win-win manner
- Concern for documentation—able to document changes and solutions to problems as work progresses regardless of other pressures
- Professional integrity—strives to fulfill commitments, even when inconvenient, in a scrupulous manner
- Technical curiosity—is challenged by problems that do not have a standard solution, reviews current methods to find innovative solutions, and takes an active role in solving unfamiliar problems by listening to others to learn new concepts

- Testing hypothesis on others—asks colleagues' opinions regarding a planned approach to a problem, listens to others' opinions and criticisms without becoming defensive or embarrassed, and acts on recommendations.

This list covers a few of the more important attributes required, but is by no means comprehensive.

3.6.6 Staff competence and motivation

Initial staff selection needs to be based on proven competence and motivation, and training peculiar to the turnaround needs to be initiated. Generic training includes permit to work, safety, etc. This generally takes place in the preexecution phase and is discussed further in Section 6.6.1. However, highly specialized training for planning staff is required up to 18 months prior to the turnaround. Appendix C is the starting point for this training.

Box 3.11 gives an example of a change in **management competence** and **contracting strategy** as a result of change in ownership.

Empowering the team boosts motivation. The philosophy statement referred to in Section 3.1.2 is a key tool for empowerment.

Box 3.11: Competence

A refinery changed ownership in the year 2000. Shortly afterward, a crude unit turnaround was being undertaken with a planned duration of 40 days. At the time, Solomon Associates benchmarking estimated that the turnaround duration for this type and size of crude unit was 30 days.

The new owners raised the question as to whether the duration could not be reduced to 30 days, and the answer was: "We have always done it in this time and cannot do better." (The turnaround contract involved just a single turnaround contractor based on a bill of quantities, which does not motivate a contractor to reduce duration. On the contrary, a contractor is motivated to maximize the amount of work.)

However, a search of old records found that under the previous ownership this particular crude unit turnaround had, in fact, been completed in 30 days. Even when confronted with this fact, the refinery leadership was not willing to change.

Lesson learned

The new management should have been led through a paradigm shift using a facilitated change management process by an independent facilitator.

It was clear that the following was not evident and was required:
1. Adequate full-time turnaround planning staff with critical path modeling experience
2. A structured scope challenge process
3. An integrated critical path model that had been archived after each turnaround and copied and improved on for the next turnaround
4. A contracting strategy to employ and motivate very experienced turnaround contractors to complete the turnaround in the required duration.

Shell Global Solutions (SGS) could have possibly offered the service as an independent facilitator as was done in the case described in Box 5.5: "Scope Challenge—Gas Plant Turnaround."

Comment

Resistance to change is greatest when people think they are performing well but, in fact, are not. The leadership of this refinery clearly thought they were doing fine and were not willing to change. Only when a leader understands that the competition is doing better by taking a different approach can things start to change.

3.6.7 Trade resource planning

Very early resource planning is required as the need for large numbers of skilled workers over the short period of a turnaround affects other industries in the area or country.

Industries in countries or areas often collaborate to ensure turnarounds don't overlap. An example is shown in Box 3.12, which indicates both dates of various turnarounds as well as the peak workforce requirements for different companies in the region.

Preliminary resourcing estimates are determined from the CPM model of the previous turnaround on the same plant, and its productivity factor (see Section 5.6.5).

Example

Resourcing in the Middle East, particularly in the oil and gas industry, requires early planning as resources are sourced from countries as diverse as India, Pakistan, and the Philippines.

3.6.8 Booking of specialist services and equipment

Simply put, if essential materials are not available, the work cannot be done. Early evaluation of required services and tools would highlight the necessity of early booking of these requirements.

Services
Prompt booking of specialist services includes:

1. Fluidic catalytic cracking unit (FCCU) slide valve experts
2. Gas turbine experts
3. Catalyst change experts.

Equipment
Specialist equipment that requires timely arranging includes:

1. Cranes for heavy lifts
2. Postweld heat treatment (PWHT) rigs
3. Ultra-high pressure (UHP) water cleaning equipment for hard-to-clean exchangers, etc.

Example

The use of PWHT rigs or UHP water cleaning equipment may be required for critical path activities. Inadequate numbers of these may affect the critical path.

Special heavy lift cranes are often in short supply and not always available in the region. They therefore need to be booked well in advance and transported timeously to the site, sometimes from other countries.

Box 3.13 describes the disposal of a specialized crane before it became apparent that it was required for very particular lifts during turnarounds.

Appendix H lists expert contractors.

Box 3.12: Canadian tar sands collaboration

2007 Major Turnarounds/Shutdowns - TENTATIVE SCHEDULE

> **Box 3.13: Specialist crane**
>
> Cape Town Refinery had an American lattice boom crawler crane that lay disused in the yard for many years.
>
> New management decided to dispose of the crane, thinking that it was redundant equipment from a refinery expansion project. However, this type of crane was required for very specific lifts to access the highest points of the plant during turnarounds. (Apparently the crane was only used every 4 years during major turnarounds.) The refinery felt obliged to buy the crane back from the purchaser at an elevated price in time for the next major turnaround.
>
> **Comments**
>
> Today, specialist heavy lift cranes are readily available for hire in most parts of the world. It might have been far cheaper to hire this sort of crane rather than maintaining and storing it for extended periods when not in use. Having said that, large pieces of equipment can be difficult to transport.
>
> In the 1990s, Bahrain Petroleum Company had to hire a large crane from Dubai and relocate it by ship for a special fluidic catalytic cracking unit turnaround project (see Box 4.9).

3.7 Communication

3.7.0 Introduction and relevant PM processes

In phase 1 the communications framework is established or updated.

Related standard processes are:

PMBOK	ISO 21500
10.1a Plan communications management	4.3.38 Plan communications

See Appendix A7 for details.

3.7.1 Communication framework

The communications framework needs to be established at a very early stage and communication methods need to be refined. These include:

- Formal meetings for making decisions, deciding actions, assigning actions, and ensuring implementation, including addressing HSE issues
- Informal meetings for problem solving, team building, etc.
- Briefings for passing on information agreed by management
- Audio visuals for training purposes and packaged presentations such as "rules for access to site," etc.
- Written material including project/contract/job packages, e-mails, published awareness articles, contractual correspondence, etc.
- Vocal one-on-one or one-on-many including face-to-face, telephone, and radio communication
- Recording of data electronically in document and photographic (still and video) form.

Groupings of formal meetings are proposed as shown in Box 3.14.

Box 3.14: Proposed meetings

1. Integration

a. Standards and procedures

Purpose

To ensure the turnaround is executed to required standards and procedures.

b. Shutdown and start-up

Purpose

To integrate SDSU plans with master operating plan.

2. Scope

c. Work list meeting

Purpose

To gather together, and validate, the work requests generated by a large number of people in various departments, creating an approved work list for the turnaround.

d. Major task review meeting

Purpose

To ensure that large, complex, or hazardous tasks are given the due consideration that their importance merits, so that a relevant specification for the task may be produced (see Box 4.2 for examples).

e. Inspection review meeting

Purpose

To ensure that all necessary statutory and internally generated inspection requirements are identified and defined and met via the most relevant techniques.

f. Project work review meeting

Purpose

To ensure that any project work, which is normally planned by the project engineering department, is properly integrated into the turnaround schedule so that all conflicts are resolved and requirements met (see Section 5.1.2 for further discussion).

3. Schedule

g. Daily work progress (execution phase)

Purpose

To integrate and resolve all turnaround issues every 24 h.

5. Quality

h. Quality review

Purpose

To ensure quality targets are met.

6. Resources

i. Resourcing

Purpose

To ensure adequate resourcing

7. Communication

j. Site logistics

Purpose

To ensure site is organized during the event

8. Risk

k. Safety review

Purpose

To develop and implement a safe system for work

Box 3.14: Proposed meetings—cont'd

9. Procurement
 l. Contract review
 Purpose
 To ensure all contracted work is specified and managed.
 m. Spares and tools review
 Purpose
 Define items required for the event

10. Stakeholders
 n. Client/sponsor progress meetings
 Purpose
 Appraise the client of progress and any issues inhibiting meeting the agreed turnaround targets.

Circulation lists for each type of correspondence need to be established. Telephone contact lists for all key staff, contractor management, sponsors, etc. need to be displayed in prominent positions on site.

Organization breakdown structures are part of the model for the execution phase (see Sections 4.3.1, 5.3.1, and Appendix C2).

Responsibility matrices (RACI charts) are required for numerous decisions. (RACI: review, action, consult, and information). Many are embedded in the company enterprise resource planning system.

Example

Tendering and Contracting Process in SAP.

However, they may need to be created for, and embedded in, those processes that are not computerized.

3.7.2 Problem solving

Throughout the turnaround, complications will crop up, many of which may not have occurred previously. Various problem solving techniques can be implemented to find a resolution, the most common one being **brainstorming,** which involves getting a large number of ideas from a group of people in a short time. Alternatively, de Bono's "6 hats"[1] is also useful. During turnaround execution, problems typically need to be addressed within hours as work progress is measured by the hour.

Problem solving also includes conflict resolution. Conflicts cannot be permitted to fester. Should they not be resolved quickly at the lower level, then referral to higher levels of management for timely resolution is vital (see also Section 7.7.6).

3.7.3 Meeting management: training example[2]

John Cleese's amusing training video "Meetings bloody meetings" from the 1970s remains current, giving an excellent guide for shorter and more productive meetings. The key points are:

1. Plan
 a) Be clear in your mind of the precise objectives of the meeting
 b) Be clear why you need the meeting and list subjects to be covered

2. Inform
   a) Make sure everyone knows exactly what is being discussed, why, and what you want from the discussion
   b) Anticipate what information is needed
   c) Make sure participants are there
3. Prepare
   a) Prepare the logical sequence of items to be discussed
   b) Prepare time allocation of each item based on its importance and not its urgency
4. Structure and control
   a) First present evidence, then an interpretation, and then the proposed action
   b) Prevent people jumping ahead or going back over old ground .
5. Summarize and record
   a) Summarize all decisions and immediately record with the name of person responsible for any action
   b) Record and share minutes of meeting with all participants.

3.8 Risk (including HSE issues)

3.8.0 Introduction and relevant PM processes

In phase 1, risk related to the turnaround is introduced and the international standard for risk management is outlined. Risk mitigation includes an audit at the end of each phase, starting with an audit at the end of phase 1.

Related standard processes are:

PMBOK	ISO 21500	ISO 31000 risk management
11.1a Plan risk management	—	1. Establish the context
—	—	2. Communicate and consult

See Appendix A8 for details.

3.8.1 Risk overview

Planning for possible high risks is required, **especially if extensive engineering work is to take place during the turnaround.** Additional cost and duration contingencies might be needed in the budget proposal (see Box 3.8 and Box 4.9 *Method 2*). Risk is covered in greater depth in Sections 4.8 and 5.8.

3.8.2 ISO 31000 risk management[3]

ISO 31000 is the international standard for risk management.

The key requirements of the process are to:

1. Establish the context
   - Identify responsibility and reporting lines for risk
2. Communication and consultation
   - Determine your risk appetite
3. Risk assessment
   - List risks within categories (suggestions: strategic, operational, financial, compliance, environmental)
   - Rank risks by, for example, using a matrix diagram
   - Define appropriate action for risks according to the 4 Ts: terminate, transfer, treat, tolerate
4. Risk treatment
   - Take action
5. Monitoring and review
   - Monitor regularly through improved risk reporting
   - Learn the lessons and feed back into the process.

The process is depicted as shown in Fig. 3.2.

Further details are discussed in Appendix A8.

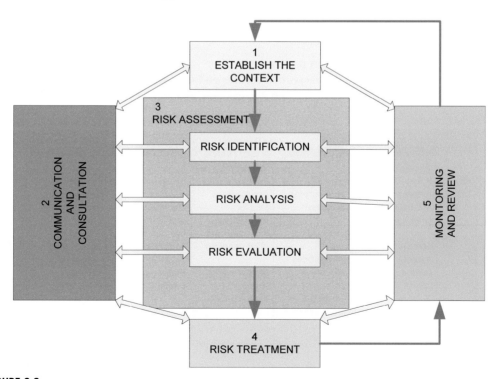

FIGURE 3.2

ISO 31000 process.

3.8.3 Turnaround risk profile

This profile is applicable to any process plant containing hydrocarbons that operate under pressure.

During normal operation, without plant upsets, the risk profile remains steady. However, risk increases with the opening of the pressurized hydrocarbon containing vessels, equipment, and pipelines. Once all of the hydrocarbon-containing vessels, equipment, and pipelines are gas free, the risk profile reduces. The highest risk profile is at start-up when hydrocarbons are reintroduced into the pressure system, which is when the potential for leaks is highest. Flawless start-up (**no incidents**, **no leaks**) is an important key performance indicator (KPI).

A turnaround risk profile is identified as per Fig. 3.3.

3.8.4 Management risks to be mitigated from the onset

The following problems are typical in turnarounds. To minimize negative issues, these need to be addressed at the start by attempting to preempt likely risks that may disrupt the event and prevent the team from achieving their agreed objectives.

Knowledge area 2: scope
Unexpected technical problems generating extra work

Detailed analysis of the more risky method statements by experienced construction and engineering staff during the conceptual development and detailed planning phases needs to be carried out to mitigate unexpected technical problems.

Uncovering work additional to the original plan

Reliable historical maintenance and inspection records and the use of online observations and predictive methods such as thickness gauging, infrared cameras, and process performance data need to be

		SHUTDOWN PERIOD			
	NORMAL OPERATIONS	**PREPARATION**	**MAINTENANCE WINDOW**	**START-UP**	**NORMAL OPERATIONS**
Risk Drivers	Pressurized hydrocarbons and hazardous chemicals Some risk of hydrocarbon release	Isolating/depressurizing equipment Continued risk from hydrocarbons and work Risk slightly elevated than normal operations	Running multiple simultaneous jobs for maintenance, inspection and plant upgrades Minimized risk from hydrocarbons	Recommissioning Increased risk profile from hydrocarbons and work **Highest risk profile**	Pressurized hydrocarbons and hazardous chemicals Some risk of hydrocarbon release.

FIGURE 3.3

Turnaround risk profile.

maximized by experienced staff to reduce the risk of surprises. A structured change control process is required for use after the scope has been frozen.

Knowledge area 3: schedule
Time lost due to disorganized plant shutdown
The key is to have a phased approach to planning and executing the turnaround with regular audits at the end of each phase. The exact sequence of turnaround and start-up needs to be detailed in the critical path model and must be communicated to all affected persons so that they are ready to carry out their duties the moment they are required to do so.

Knowledge area 4: cost
Poor cost control during the execution phase as a result of a flood of extra work
The recording of all approved changes must be structured and easy to implement. Strict control of scope change and additional work is required to prevent unauthorized work and the potential for disputes after the turnaround.

Knowledge area 5: quality
Bad workmanship
It is essential to establish all required quality control procedures and skills registers, as well as enforcement of required standards using inspections, tests, reviews, and audits.

Knowledge area 6: resources
Control team losing direction or motivation
It is critical to keep the philosophy/statement of commitment document, which has been signed by the key players, in mind to maintain focus. Fatigue is inevitable and needs to be managed by all team leaders to prevent a drop in productivity or the likelihood of an unsafe act.

Knowledge area 7: communication
Industrial action/bad industrial relations
This can be highly disruptive to the outcome of the turnaround. Issues need to be put on the table long before the start of the turnaround and contingency plans established if the risk of industrial action during the turnaround is high (see Box 7.7).

Time lost due to disorganized plant start-up
As noted in *Knowledge area 3* above, the critical path model for sequenced start-up is essential, as is detailed communication between the turnaround team, operations, and online maintenance staff. Check lists are essential, especially if a new plant is being commissioned as part of the start-up (see Appendix D).

Knowledge area 8: risk
HSE risks such as injuries, damage to property, or pollution
Clear mitigation steps are required such as an active safety program, including the use of "stop cards" for anyone seen carrying out an unsafe act; clear start-up procedures and checklists; and pollution mitigation processes, especially during shutting down and starting up (see Section 6.10.2).

Example

Petronas has a procedure for recording UAUC (Unsafe Act Unsafe Condition) online, on a daily basis, among contractors and staff.

Knowledge area 9: procurement
Client/contractor or contractor/contractor disputes

Ideally, cordial long-term relationships have been established between the client and primary contractors and between the primary contractors and their subcontractors. Contractor performance reviews after previous turnarounds should highlight any disputes and how they were resolved so as to prevent reoccurrences in the upcoming turnaround. New contractors need to be assessed in the bidding stage regarding their approach to seeking "legitimate" compensation. The contract document also has to be clear with respect to scope and provide instructions as to how to resolve disputes quickly and efficiently.

Knowledge area 10: stakeholder
Change of intent by management

The establishment of a turnaround project charter is essential to ensure full agreement between management and the project team to prevent any subsequent fundamental changes to the project.

An effective turnaround project manager will prevent a problem from "festering" and would immediately concentrate on solving the problem. Thus active communication of the nature of the problem and the solution needs to be communicated to affected persons promptly (see also Section 3.7.2).

3.8.5 Critical and near-critical activity risk assessment

A review of critical and near-critical activities needs to be assessed for risk after finalization of the CPM model, but just prior to actually freezing the model to establish the baseline. Probability of achieving the durations that have been set should be evaluated. PERT is a method that could be used as described in Appendix C6.

3.8.6 Scope growth risk assessment

In the past, wear damage was an unknown until the plant was actually opened up for inspection. Today, this risk is much reduced due to the use of predictive inspection tools (mentioned in Section 4.2.2) and the statistical approach used for duration determination (see Section 5.3.2).

High-Risk Example

Poor corrosion management identified by rapid deterioration of corrosion coupons as well as graphs showing changes in pH and induced current.

3.8.7 Procurement risks

Procurement risk mitigation is critical for a successful turnaround. This is discussed in detail in Section 3.9.4.

3.8.8 Audits and reviews

To ensure efficient execution of the turnaround, auditing is required at the end of various phases of development. This ensures that the turnaround planning process receives proper priority. The specific intervals are decided by the decision executive and the core team, but the following is recommended:

- At the end of phase 1—conceptual development
- At the end of phase 2—turnaround work development
- **At the end of phase 3—detailed planning (highly recommended)**
- At the end of phase 4—pre-execution work
- **At the end of phase 6—post-turnaround**—before the turnaround report is finalized so that its recommendations can be included in this report **(highly recommended)**

The audit at the end of phase 3 should include a work scope peer review by an independent peer review team or a shell-type scope challenge workshop comprising members of the turnaround team as well as independent members (see Section 5.8.3 for details).

The following are guidelines for conducting audits:

- Audits should be carried out by knowledgeable persons, who are not directly involved with the turnaround
- Audits shall review progress relative to the planned progress on the milestone schedule
- A brief written report should be produced for top management
- The final audit shall evaluate the team's results relative to the philosophy and commitment statements as well as the planning and execution processes. It should also ensure proper documentation of lessons learned and recommendations for improvement in the next turnaround.

3.9 Procurement
3.9.0 Introduction and relevant PM processes

The initial procurement planning takes place in phase 1. This includes identification and ordering of long-delivery items and might require provisional fund approval prior to the annual BP&B approval.

Alternative contracting strategies are studied and an appropriate strategy, which includes incentivization of the contractors, is approved.

Relevant standard processes are:

PMBOK	ISO 21500
12.1a Plan procurement management	4.3.35 Plan procurements

See Appendix A9 for details.

3.9.1 Procurement planning overview

Very early procurement planning is required with respect to the strategy for managing the turnaround and allocation of work within the turnaround. Contractor incentivization can be a major factor in reducing the turnaround duration at little extra cost. The **budget proposal** needs to accommodate the contracting strategy, especially if management is to be outsourced.

Prescreening of suitable contractors is crucial since contractors without good track records on turnarounds can raise the risk of duration extension, quality degradation, and cost increase. Checking contractor evaluations from previous turnarounds is critical for the prescreening of suitable contractors.

Prescreened, possible and suitable contractors need to be contacted and their availability for the time of the proposed turnaround ascertained.

3.9.2 Materials management

In order to ensure strict quality control, materials and spares are typically free issue from the client to contractors. The client has the required materials specifications for purchase embedded in the purchasing module of their ERM system. An example is given in Box 3.15.

Materials management is typically handled within the materials module of the company ERM system (SAP, Oracle, etc.) All purchasing of materials (direct purchase and warehouse stock items) are managed within this system. (Linkage with the maintenance management module (CMMS) is based on the master equipment database.) Thus each item of equipment would have requisite spares attached to it in the system, with a corresponding warehouse stock reference number. As turnaround planning progresses, stock items are reserved for the turnaround and are thus not available for routine on-stream maintenance.

Box 3.15: Materials quality control example

International Oil Companies (IOCs) have developed internal design engineering practices (DEPs) linked to national or international standards over many years. These have associated Standard Drawings and Materials and Equipment Standards and Codes (MESCs).

Example: piping classes
The DEP contains predesigned piping systems for a variety of services. These have a direct link to the MESC materials catalog for purchasing.

A collection of standardized piping components (caps, elbows, tees, reducers, branch fittings, check valves, gate valves, globe valves, ball valves, and butterfly valves) that are compatible and suitable for a defined service at stated pressure and temperature limits are specified for each class of pipe.

The use of the MESC materials catalog for making purchases results in a large reduction in engineering and procurement effort, and also improves integrity/quality control, promotes company-wide standardization, and reduces the level and variety of spares holding.

There are also instances, however, where spares and materials are more conveniently supplied by the contractor.

Examples

1. *A turbine vendor contracted to service and repair a turbine and therefore supplying all required spares.*
2. *A thermal insulation contractor supplying and installing thermal insulation and cladding.*

Fig. 3.4 shows the materials requisitioning process.

3.9.3 Long-lead items

Long-lead items have to be identified at this stage and a preliminary expenditure request may be required to trigger purchase of these items. A cancellation clause might be required for those items where the required specification and quantity cannot be determined or may be changed.

Example

Inspection has decided to replace a number of furnace tubes in a particular furnace. However, they are not sure how many to replace and if they need to change the materials specification. Initially, an order would be placed for a number in excess of what is estimated to be required and with the same specification as before.

Analysis of the preliminary work list would identify long-lead items such as:

- Rotary equipment spares
- Large or special valves, pumps, and turbines
- Nonstandard items such as FCC regenerator Y-pieces
- Exotic materials such as alloy piping and furnace tubes
- Large quantities of furnace tubes, column fill, etc.
- Replacement columns, vessels, and heat exchangers.

Box 3.16 describes a situation where a long-lead order was missed, which resulted in major expenditure to get the materials on site in time for the turnaround.

Box 3.17 gives an example of an expediting spreadsheet.

3.9.4 Procurement risk and performance

The tendering process is discussed in Section 5.9.2.

The procurement of materials (feed stock, process, maintenance, etc.) and services (maintenance, turnaround, logistics, etc.) are major contributors to risk control and performance improvement.

Risk mitigation related to the type of contract is very important. Table 3.3 illustrates the risk spectrum for selection of the most appropriate contract for a specific requirement. Risk is often transferred to the contractor, depending on the type of contract.

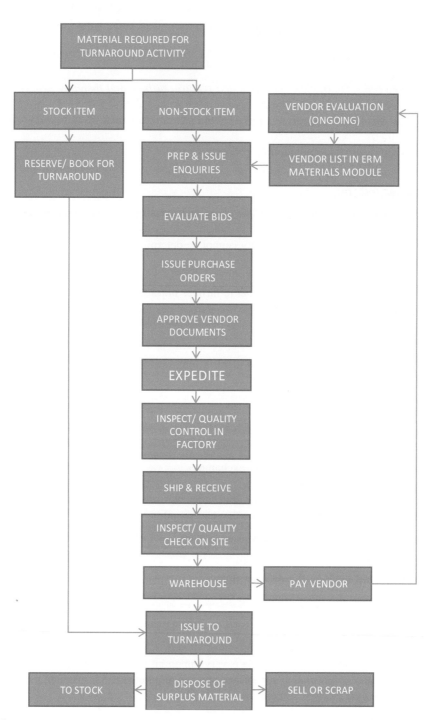

FIGURE 3.4

Materials requisitioning process.

> **Box 3.16: Long-lead item "forgotten"**
>
> Cape Town Refinery's turnaround planning team initiated the preparation process for the following year's turnaround. Long-lead items were identified, among them being furnace tubes.
>
> A materials request for furnace tubes was subsequently entered into the shutdown planning acceptance process (SAP) purchasing system for approval and delivery well before the turnaround.
> However, the approver failed to see the item queued for his approval until it was too late to sea freight the furnace tubes from Italy to South Africa in time for the turnaround. They had to be airfreighted, at tremendous cost, to reach the site in time.
>
> **Comment**
>
> At the time, all staff were going through a learning process with regard to SAP implementation. The approver was obviously not aware of this item hanging in the system awaiting this approval. In addition, there was no formal expediting process in place that might have highlighted the oversight.
>
> An expediting system needs to be established for all materials that need to be purchased for each turnaround.

Example 1

A lump sum or fixed price contract will ensure that the price will not be adjusted if the client does not issue variation orders. The scope needs to be very clearly defined before going to tender. The **advantage** *to the client is that the risk is passed to the contractor. However, a potential* **disadvantage** *is that, if variation orders are issued, part of the risk moves back to the client, and the cost normally goes up. Contractors are incentivized to complete the work quickly and efficiently.*

Example 2

A cost reimbursable contract price is determined by the final measurement of work. The contract is determined by a bill of quantities (BoQ) with estimates of work to be done and the associated rates for each piece of work. The **advantage** *to the client is that work can be fast-tracked, as the design does not have to be complete when awarding the contract. The* **disadvantage** *to the client is that there is a high risk of the cost being greater than initially estimated.*

3.9.5 Contracting strategies

Service contracts

There are four common types of contracts:

1. Lump sum
2. Unit price
3. Target cost
4. Cost reimbursable

Lump sum contracts are ordinarily used for activities like valve refurbishment where the total number of valves is known. Other examples of lump sum contracts are the cleaning, inspection, and testing of heat exchangers. The contractor may be able to take them offsite and set up a production line operation in their workshops. The risk then passes to the contractor. This sort of contract can be enhanced with an incentive bonus/penalty to improve productivity. Any engineering project work that is normally well defined can be awarded on a fixed price (lump sum) basis.

Box 3.17: Turnaround material status spreadsheet example

SHUTDOWN MATERIAL STATUS (Shutdown 2009)													
Date:	29-Apr-2017					Shutdown Date: 1-Jan-2009	Refresh		Status Report No:	SDSR-001-08			

Table columns: Location & Item Description | Material | Is it Critical Item? | Requisition Order No. | Purchase Requisition Date | Purchase Order No | Purchase Order Date | Value Range (QR) | Purchase Order Processing Time (Material value dependant. In days) | Material Production Time (Material dependant. In days) | Delivery Time (Days) | Total time to receive material (Days) | P.R./P.O. date to Shutdown Material Procurement Target (01-Jan-2009) Available Days | On-Time Flag

Location & Item Description	Material	Is it Critical Item?	Requisition Order No.	Purchase Requisition Date	Purchase Order No	Purchase Order Date	Value Range (QR)	PO Proc Time	Material Prod Time	Delivery Time	Total time to receive	Available Days	On-Time Flag
U41	FURNACE TUBES, ALLOY STEEL	YES	SDR-R13-08	30-Oct-2007	SDO-R938-08	12-Mar-2008	50,001 - 150,000	42	210	56	308	295	
U41	FURNACE TUBES, ALLOY STEEL	YES	SDR-R28-08	23-Oct-2007		12-Mar-2008	50,001 - 150,000	42	210	56	308	295	
U51	BOILER/SUPERHEATER TUBES, CS	YES	SDR-R45-08	10-Feb-2008			> 3 Million	112	105	56	374	326	
U21	HEAT EXCHANGER TUBES, CS	NO					5001 - 50,000	28	105	56	290	295	
U61	BUTT WELD FITTINGS, CS/LAS/NA/SS	YES					5001 - 50,000	28	112	56	297	295	
								0	0	56	157	568	

Material could be procured on time
Material could not be procured on time - Avail other options to procure material
Could lead to insufficient time to procure material if purchase requisition has made on time
Could lead to insufficient time to procure material - Critical Item

The above spreadsheet was developed as follows:

Calculation

Total Delay=Purchase Requisition processing time +Purchase Order processing time (Material Value Dependant) +Material Production Time (Material Dependant) +Delivery Time (Assumes maximum time of 8 weeks)

Material Production time & Purchase Order Processing time data are taken from the company materials system and is stored locally in a sheet of the application. This sheet needs to be updated manually for the materials that are not listed.

How it works

1. Create a status report file
2. Add a required material by choosing from   the drop down list
3. Indicate whether it is a critical item by choosing Yes/No from dropdown options
4. Select the value range
Above 3 entries are mandatory
5. If a purchase request has been made enter Purchase Requisition date &   P.R. number
6. Based on these available data it calculates the delay time and shows the status in corresponding color Red/Green. If the available time is not enough to procure the material it gives RED flag
7. If P.R. Date has not entered the software gives indication based on current date, but can give different color Red/Green/Orange /Yellow also based on criticality of the material

- **RED.** Material can't procure on time Use alternate methods to procure the material e.g.: Air Cargo
- **GREEN.** Material could procure on time based on the calculated delays
- **ORANGE.**   P.R. Not Created, also it is a critical item. If P.R. has not created on time it can lead to Red
- **YELLOW.**   P.R. Not Created, Not a critical item, But it can lead to red if P.R. has not created on time.

If P.R. has not made and if software finds, "material can't procure on time even if P.R. is made TODAY", then it indicated RED

If P.R. has not made and if software finds, the material can be procure on time if P.R. is made TODAY, but it indicates ORANGE/YELLOW based on criticality of Material, because it warns if P.R. not made today it can lead to RED.

Note: This spreadsheet is an example for purchase of items from established vendors. Times were averages from SAP Materials Module historical records.

Table 3.3 Contracting risk spectrum.

No	Issue	Lump sum	Unit price (BoQ)	Target cost	Cost reimbursable	Target man-hours (for turnarounds)
1	**Financial objectives of client and contractor**	Different but reasonably independent	Different and in potential conflict	Considerable harmony. Reduction of actual cost is a common objective provided cost remains within the incentive region	Both, based on actual cost but in potential conflict	Considerable harmony. Reduction of actual man-hours is a common objective
2	**Contractor's involvement in design**	Excluded if competitive price based on full design and specification	Usually excluded	Contractor encouraged to contribute ideas for reducing cost	Contractor maybe appointed for design input prior to execution	Contractor intimately involved with CPM modeling to determine estimated man-hours for the project
3	**Client involvement in management of execution**	Excluded	Virtually excluded	Possible through joint planning	Should be active involvement	Jointly involved with contractor
4	**Claims resolution**	Very difficult, no basis for $ evaluation	Difficult, only limited basis for $ evaluation	Potentially easy, based on actual costs. Contract needs careful drafting	Unnecessary except for fee adjustment. Usually relatively easy	Unnecessary as actual man-hours plus bonus (based on a formula) basis for payment
5	**Forecast final cost at time of bid**	Known, except for unknown claims and changes	Uncertain, depending on quantity variations and unknown claims and changes	Uncertain. Target cost usually increased by changes, but effective joint management and efficient working can reduce final cost below an original realistic budget	Unknown	Estimate based on previous CPM model for turnaround of the same plant
6	**Payment for cost of risk events**	Depending on contract terms, undisclosed contingency, if any, in contractor's bid. Otherwise by claim and negotiation	Depending on contract terms, undisclosed contingency, if any, in contractor's bid. Otherwise by claim and negotiation	Payment of actual cost of dealing with risks as they occur and target adjusted accordingly	Payment of actual costs	Payment for actual man-hours plus overheads

Adapted from Wideman RM. Project and program risk management: a guide to managing project risks and opportunities. PMI; 1992[6].

Unit price contracts have been successfully used for scaffolding, which puts the onus on the contractor to determine the right materials and labor for a particular scaffold arrangement. It has also been used in other measurable jobs such as valve refurbishment. However, it is not suitable for major turnaround work as the contractor is incentivized by an increase in amount of work with possible deterioration in quality (see Box 3.11).

Target cost contracts are more easily understood when considering the concept of partnering. Partnering requires bringing the contractor and client together to develop mutual goals and mechanisms for solving problems.

Cost reimbursable contracts tend to be used more often since it is generally difficult to determine the exact scope of work for large portions of a turnaround. Payment is based on the hours worked by the contractor's staff. Regrettably, this puts most of the risk on the client and also makes it difficult to motivate contract staff whose main incentive is to maximize the number of hours that they work in order to increase their earnings.

Target cost and reimbursable contracts can be further refined by, for example, the contractors' and client's planners conferring months before the turnaround to build the critical path network and resource histograms together. A target total number of man-hours can be determined and used as a basis for an incentive type contract. This type of contract has been executed very successfully in major refinery turnarounds around the world.

Incentive-type contracts require a mature approach to contracting based on a positive long-term relationship between the client and the contractor. Incentive factors, other than man-hours, can be based on quality reworks, safety incidents, etc. The incentive has to be positioned on a win-win contract basis with common objectives such as duration reduction, and minimum quality reworks and safety incidents.

It is important to note that incentive-type contracts have to be simple to administer. Thus tracking of the incentive values needs to be automated as far as possible.

Materials purchasing contracts

Materials purchasing contracts may require a different approach to service contracts. The vendor's performance over a number of years should be recorded in the materials module of the company's ERM system. Items like actual delivery relative to promised delivery are critical in assessing the probability of meeting target dates. This data should be reviewed before award of contract. Long-delivery items have to be expedited by the client or his representative, to ensure manufacture is progressing to schedule and target dates for delivery are met. A detailed purchasing plan is thus necessary (see also Section 3.9.3).

Term contracts for a specific period are useful for on-demand supply of materials, thus reducing warehousing costs. The contractual agreement requires a clause for response time for delivery of materials. Contracts include supply of relief valve spares, mechanical seals (possibly including refurbishment of replaced seals), and modular exchange of standard gas turbines (with refurbishment of the replaced turbine).

3.9.6 Division of labor

A client generally does not want to put all its "eggs in one basket." Since several contractors would be involved, resources could be used from a similar contractor to get back on track, should one contractor fall behind.

Box 3.18: Division of labor example

Bahrain Petroleum Company (Bapco) has a 250,000 barrel per day refinery and its own dedicated power generation and desalination plant. Turnarounds continue throughout the year where process units/complexes are independently taken out of service. The refinery power turbines and boilers, as well as the desalination units, are also independently taken out of service for planned outages.

Bapco has traditionally awarded long-term contracts for turnarounds in the following categories:
1. Columns and vessels
2. Heaters
3. Exchangers
4. Lines and fittings
5. Insulation/refractory
6. Electrical
7. Instrument
8. Rigging
9. Staging
10. Civil

The same management team from each contractor is involved with each turnaround, and they are therefore thoroughly knowledgeable about all aspects of the turnaround for which they are responsible.

An example of awarding of contracts by type of equipment is given in Box 3.18.

3.9.7 Approaches to improve procurement performance

The following are examples of useful methods to improve procurement performance with the ALARP risk approach.

Improved contracts system KPIs

KPIs need to be linked to real improvement in the overall management of the business.

> *Example*
>
> *Critical long-term deliveries of equipment and materials for arrival on a specific critical date should be monitored individually. This is vital for a scheduled turnaround.*

Scope definition

Clear scope definition is required for a tender to be priced. If this is not possible, then a BoQ contract may be more suitable. The changing risk profile is shown in Table 3.3.

Competence of contractors

Major national oil companies and international oil companies (IOCs) classify the competence of their contractors prior to bidding. This is based on the contractors' experience in certain types of work, the complexity of the work, and the size of contract. Only those in a certain classification are invited to bid for the work.

Prequalification

In addition to competence classification, prequalification might also be required for activities that are of higher risk, and thus the experience of the contractor in the specific required field is crucial (see Box 2.5).

Eliminate postbid price negotiations

International practice for formal bidding does not allow for postbid price negotiation. Regrettably, this practice removes the transparency and fairness entailed in a formal sealed bid process. With this practice, contractors are often compelled to reduce their profit margins to a high-risk situation in order to win the contract, often resulting in poor performance.

Contracting strategies

Strategies need to be tailor-made for the situation. Incentive schemes for turnaround contractors help, focusing on completing the turnaround on target and in budget.

> *Example*
>
> *A catalytic cracker (FCCU) shutdown could have a lost opportunity cost (lost profit) well in excess of $100,000 per day.*

3.10 Stakeholder management
3.10.0 Introduction and relevant PM processes

Stakeholders are identified in phase 1. Buy-in from the primary stakeholder, the decision executive/sponsor, is critical, as this is their agreement, with milestones, for the turnaround.

Related standard processes are:

PMBOK	ISO 21500
13.1 Identify stakeholders	4.3.9 Identify stakeholders
13.2 Plan stakeholder engagement	-

See Appendix A10 for details.

3.10.1 Stakeholder identification and roles
Definition: project stakeholders

Individuals and organizations who are actively involved in the project, or whose interests may be affected as a result of project execution or project completion.

The following primary stakeholders need to be identified

1. Physical asset owner (shareholder)
2. Asset operating and maintenance company (sponsor)
3. Suitable contractors for carrying out turnaround work
4. Regulatory authorities requiring notification of the turnaround (see Section 3.10.2)

5. Suppliers of raw materials
6. Customers of products.

The contractor—union relationships possibly also need attention at this stage to avoid issues that might escalate when the turnaround execution is in progress (see Box 7.7).

The sponsor establishes a steering committee or policy team, which is drawn from managers and engineers who fulfill at least one of the following criteria:

1. They are stakeholders who are directly affected by the turnaround
2. They provide the funds to pay for the turnaround
3. They have the authority to make decisions concerning the turnaround.

However, the key players referred to in this text are the decision executive (sponsor) and the turnaround project manager who manages the core team.

Fig. 3.5 depicts the turnaround stakeholders and relationships.

3.10.2 Getting buy-in
Decision executive
At the end of phase 1, the core team will have produced an OOM estimate and a provisional critical path analysis giving total duration of the turnaround. If this is a repeat of previous turnarounds, the estimate and duration should be much more accurate, providing there is no major capital work. The *decision executive* gives guidance on the proposed budget and duration and concurs once the team has taken note of the concerns of the decision executive. Details of how the estimate and duration are determined are discussed in Sections 3.3 and 3.4.

In addition to cost and schedule, other items presented to the decision executive and packaged as per the project charter for buy-in are:

1. Manpower forecast for development of the planning phases of the turnaround. This can be derived from the milestone schedule model.
2. A milestone schedule for the whole shutdown cycle.
3. A process shutdown and start-up sequence by process unit—a "skeleton" network diagram.
4. An audit schedule.
5. Any major risks and appropriate mitigating actions.

Regulatory authority
Buy-in is also required from regulatory authorities, especially if there is a deviation from established or agreed operations/processes/regulations.

> *Example*
>
> *In Qatar, the Environmental Regulatory Authority requires notification of a gas plant turnaround as excess (abnormal) flaring occurs during shutdown and start-up of a gas train.*

Box 3.19 describes a request to obtain exemption from established regulations.

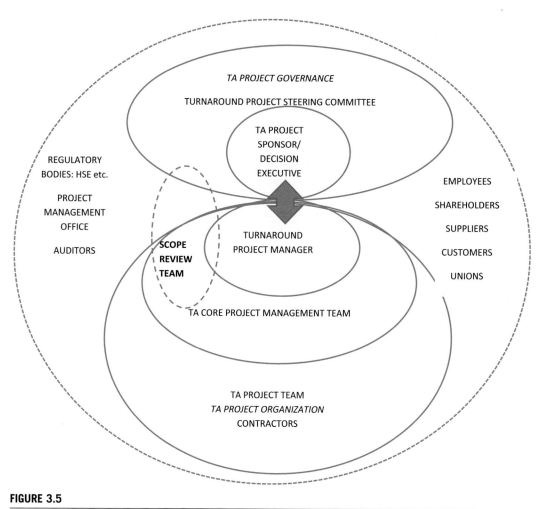

FIGURE 3.5

Stakeholder relationships.

3.10.3 Setting milestones

Turnarounds vary in complexity, cost, and manpower requirements. Timely input from all sections of the entire manufacturing operation is required for a successful turnaround, although many of those from whom input is required are not part of the team, and have other more immediate problems to solve. Thus the core team needs to establish a milestone schedule for the specific turnaround, which then becomes a "schedule of commitment" for all stakeholders.

Milestones are events that don't occupy time. For example, "start of phase 1." The milestone schedule helps to establish priorities to support the planning effort. It should be formalized and maintained by

---

Box 3.19: Visit to site by regulatory authority

Chevron Cape Town Refinery Internal Newsletter extract adapted

Visit of Department of Labor representatives to Chevron Cape Town Refinery
The inspection group was recently visited by three people from the Department of Labor. The objective of the visit was to discuss the Vessels Under Pressure Regulations exemption requests made by inspection, and to arrive at a decision as to whether or not to grant exemptions.

This sounds heavy weather, but it is! The exemptions for all companies, which previously had been granted under older versions of the government's regulations governing pressure vessels, were withdrawn by the end of 1999. These regulations are designed particularly for nonsophisticated users but apply to all.

The Department's visit lasted 4 h in all, as this is the first of numerous such discussions with industry and could have major implications on the approach to other users of pressure vessels.

Some of these regulations can be extremely restrictive to large users, and fundamentally affect the way the refinery operates. Therefore we set about convincing the Department of Labor representatives of the justification for granting these exemptions, which are required before the start of the refinery turnaround.

Outcome
The exemptions were approved by the Department of Labor.

Comment
The Department of Labor (DOL) of the South African Government enforces pressure vessel regulations in South Africa.

Chevron Cape Town refinery inspection group is a member of the South African Association of Approved Inspection Authorities (AIA).[4] AIA are those inspection bodies that are accredited by SANAS under ISO 17020 and further approved by the DOL.

ISO 17020 General Criteria for the Operation of Various Types of Bodies Performing Inspection[5] is an internationally recognized standard for the competence of inspection bodies. ISO 17020 should not be confused with ISO 9001, which is specific to quality management systems.

---

the core team leader (turnaround project manager). This document then becomes the basis for process audits.

The responsibility for getting commitment and support from other stakeholders is a core team concern. Tracking planned versus actual progress will identify problem areas so that timeous corrective action can be taken. A sample milestone schedule is shown in Box 3.7.

Long-term delivery items need to be included in the milestone schedule as these items could well delay the start of the turnaround if allowed to slip. A preliminary expenditure request might be required to place orders for these items.

3.11 Summary: phase 1—turnaround conceptual development
3.11.1 Focus areas

1. Appoint a turnaround project manager and form a core team for the particular turnaround
2. Define turnaround charter, including philosophy and goals
3. Identify work input sources

4. Review plant history and lessons learned (including HAZOP and root cause analysis results)
5. Develop work list criteria

3.11.2 Deliverables

Requirements for the **IBD** and management buy-in, including:

1. Project charter
2. Turnaround philosophy
3. Preliminary work lists
4. OOM estimates
5. Estimated turnaround duration
6. Turnaround preparation timetable (milestones)
7. Planning manpower forecast
8. Audit schedule
9. Other relevant data such as risk issues

3.11.3 Tools

1. Key job functions for core team and others
2. Turnaround planning process overview
3. Work list criteria
4. Turnaround preparation milestone schedule
5. Team building
6. Progress audit

3.11.4 Timing

12—18 months prior to the turnaround.

References

1. De Bono E. Six thinking hats. Penguin 1999.
2. Meetings bloody meetings video arts 1976.
3. ISO 31000: 2018 risk management. Available from: https://www.iso.org/iso-31000-risk-management.html .
4. South African Association of Approved Inspection Authorities. Available from: http://inspectionauthority.co.za/.
5. ISO 17020 general criteria for the operation of various types of bodies performing inspection.
6. Wideman RM. *Project and program risk management: a guide to managing project risks and opportunities.* PMI; 1992.

Phase 2 Work development

4

4.0 Outline

The **Initial Budget Decision** is the starting point for this phase and includes the agreement (project charter) as to what is to be accomplished both in the turnaround, as well as in this phase. The turnaround scope is developed to a point where a final work list can be used to develop a critical path model. Activities are categorized and alternative work methods are investigated and agreed on. Risks are assessed, quantified, and mitigated.

Issues covered in this phase are:

1. integration
2. scope
3. schedule
4. cost
5. quality
6. resources
7. communication
8. risk (including health, safety, and environmental issues)
9. procurement
10. stakeholder management

These will be examined in more detail with varied emphasis in this phase.

Specifically, Table 4.1 references PMBOK and ISO 21500 processes applicable to this phase.

The end of the phase is summarized in a proposal for a **preliminary budget decision (PBD)** based on a defined scope of work.

4.1 Integration
4.1.0 Introduction and relevant project management (PM) processes

Phase 2 entails further detailed development of the project management (PM) plan. Specific actions for this phase are carried out that include giving further details of the work required in later phases.

No	PMBOK	ISO 21500
	Table 4.1 PMBOK and ISO 21500 processes.	
1	4.2b Develop project management plan	4.3.3 Develop project plans
2	5.1b Plan scope management	–
	5.2 Collect requirements	–
	5.3 Define scope	4.3.11 Define scope
3	6.1b Plan schedule management	–
	6.2b Define activities	–
4	7.1b Plan cost management	–
	7.2b Estimate costs	4.3.25 Estimate costs
	7.3b Determine budget (for prelim budget decision)	4.3.26 Develop budget
5	8.1b Plan quality management	4.3.33 Plan quality
6	9.1b Plan resource management	
	9.2a Estimate resources	4.3.16 Estimate resources
	9.4a Develop team	4.3.18 Develop project team
7	10.1b Plan communication management	4.3.38 Plan communications
8	11.1b Plan risk management	–
	11.2b Risk identification	4.3.28 Identify risks
	11.3 Perform qualitative risk analysis	4.3.29 Assess risks
	11.4 Perform quantitative risk analysis	
	11.5 Plan risk response	–
	11.6 Implement risk responses	4.3.30 Treat risks
9	12.1b Plan procurement management	4.3.35 Plan procurements
	12.2a Conduct procurement	4.3.36 Select suppliers
10	13.3 Manage stakeholder engagement	4.3.10 Manage stakeholders

The related standard processes are:

PMBOK	ISO 21500
4.2b Develop project management plan	4.3.3 Develop project plans

See Appendix A1 for details.

4.1.1 PM plan development

The PM plan initiated in Section 3.1.3 is implemented and expanded. This covers aspects of phase 2—Work Development. All work list inputs are assembled in the preliminary work list. As the work list comes together, reviews of work list criteria, input on constructability, and reference to the turnaround philosophy keep the scope in focus.

All work list items are reviewed and assessed by the work scope review team.

As job activities are defined, critical path schedules are built on the turnaround model. Alternative methods can be computed in the model to assess the effects ("what ifs").

Box 4.1: Traffic light report for planning progress

MONTH -18: April 2017					
PHASE 2 Work Development					
Activity	Impact	PI Status	Act Status	Comment	Traffic light
ID **From SD prep sched 23/1/17**					
3 Review task lists	high	finish			
7 SOW input from all depts	high	finish			
4 Identify spares & mat from task lists	high	start			
5 Prepare PV level 4 sched	high	start			
6 Review BOQ	med	start			
11 Compile master sched & milestones	high	start			
Other					
Long lead items orders placed					
Finalise Preliminary Scope List					

Risks are evaluated using the model and other means.

A computerized maintenance management System (CMMS) enhances the decision-making process and assists with ordering of spares.

Logistics for the turnaround start to be developed.

At the end of phase 2, the updated cost estimate, schedule, and other deliverables are presented to the decision executive for guidance and concurrence.

Traffic light reports could be used for monitoring the progress of turnaround planning activities. An example is shown in Box 4.1.

4.2 Scope
4.2.0 Introduction and relevant PM processes

Scope management in phase 2 entails the collection of work requests and the screening of these requests using the criteria and process established in phase 1.

The relevant standard processes are:

PMBOK	ISO 21500
5.1b Plan scope management	—
5.2 Collect requirements	—
5.3 Define scope	4.3.11 Define scope

See Appendix A2 for details.

4.2.1 Scope development 2

Notifications and requests for work are typically entered into a CMMS by requesters, which include operations, maintenance, inspection, mechanical/electrical/instrument design engineering, and process engineering.

The maintenance planner carries out the initial screening and ensures that the wording and terminology of the request is correct. Duplicate requests from various requesters are consolidated and this results in the preliminary work list.

The work scope review team then reviews this list and embarks on a challenge process to select turnaround work based on the criteria established in phase 1. Rejected work and deferred work is returned to the requester with reasons for the rejection. This work may then be scheduled for online completion or deferred to a following turnaround. An example of the challenge process documentation is shown in Box 4.2.

Further to establishing guidelines for inclusion of work in the turnaround as listed in Section 3.2.1, all work must also follow sound engineering practices, standards, and principles of risk management.

The final work list then needs to be screened for "not in kind" work where the management of change (MOC) process is initiated. This entails engineering work, and when this is complete it is issued to the turnaround planners as a "construction turn over" package. This is then combined with routine turnaround activities to form the network diagrams as detailed in Section 4.3.2.

MOC is discussed further in Section 6.8.2.

Fig. 4.1 outlines the sequence of events from the requests for work being added, to a preliminary work list, to building activities (tasks) into the critical path method model.

4.2.2 Review and assessment methods and tools

How is the work list developed? The work scope review team must first establish guidelines for evaluation of items for inclusion in the work list. This is covered in Section 3.2.1.

The work scope review team members' ability to discuss, negotiate, and select the best items to work on during the turnaround requires a good understanding of the "big picture" as mentioned under key attributes in Section 3.6.5. Thoughtful judgment, risk analysis, and discipline are also required when choosing work list items. The efficient execution of the whole turnaround requires timely development of the **"final" work list** and adherence to this work list that essentially is the **project scope**.

Inputs for the work list are generally as follows:

1. previous turnaround reports
2. maintenance work awaiting a turnaround
3. maintenance records
4. inspection records
5. best practices
6. safety, health, environmental, and quality records
7. process performance records
8. engineering work requiring a turnaround

Box 4.2: Work list showing scope justification and approval[1]

Work Notification No.	Description	Equip No	Date created	Originator	TYPE (Capex/Opex/T&I)	Funding	Justification	Scope Approval
50006892	Fractionator 34C-1 PSV's to Atmosphere		23/09/2009	JFARHAR	CAPEX - Normal Work	Capex	Approved Capex	Yes
50006893	Ex Work - 34KM-3 EJB		23/09/2009	JFARHAR	CAPEX - Normal Work	Capex	Approved Capex	Yes
50006894	CSC Valve Improvements 34XV-014		23/09/2009	JFARHAR	CAPEX - Normal Work	Capex	Approved Capex	Yes
50006899	Replace Plant 34 Oxygen analysers		23/09/2009	JFARHAR	CAPEX - Normal Work	Capex	Approved Capex	Yes
50006902	Replace Column Packing In-Kind	A34C1	25/09/2009	LSINCLA	Turnaround - Normal Work		TSD workscope	Yes
50006903	Water Test/Adjust Liquid Distributors	A34C1	25/09/2009	LSINCLA	Turnaround - Normal Work		TSD workscope	Yes
50006904	Replace Feed Nozzles In-Kind	A34D4	25/09/2009	LSINCLA	Turnaround - Normal Work		TSD workscope	Yes
50006905	Anticipate repairs to downcomers	A34C5	25/09/2009	LSINCLA	Turnaround - Normal Work		Not justified, reject notification.	No
50006906	Anticipate repairs to downcomers	A34C6	25/09/2009	LSINCLA	Turnaround - Normal Work		Existing workscope on 34C-6, therefore approve.	Yes
50006907	Repair Flapper Valves	A34D3	25/09/2009	LSINCLA	Turnaround - Normal Work		TSD workspace	Yes
50006908	Dump/Screen/Replace Charcoal	A34D28	25/09/2009	LSINCLA	Turnaround - Normal Work		No longer required	No
50006909	Dump/Screen/Replace Charcoal	A34D29	25/09/2009	LSINCLA	Turnaround - Normal Work		No longer required	No
50006910	Renew Wye Ring	A34D4	25/09/2009	LSINCLA	Turnaround - Normal Work		TSD workscope	Yes
50006911	Renew pre-stripping steam rings	A34D4	25/09/2009	LSINCLA	Turnaround - Normal Work		TSD workscope	Yes
50006913	Inspect cyclone internals with camera	A34D4	28/09/2009	LSINCLA	Turnaround - Normal Work		TSD workscope	Yes
50006914	Siemens 34KT-2 Rotor Life Assessment	A34KT2	28/09/2009	GJCOX	Turnaround - Normal Work		Coverage for rotor overhaul costs, awaiting management approval	
50006915	Siemens 34K-2 Rotor Modifications	A34K2	28/09/2009	GJCOX	Turnaround - Normal Work		Coverage for rotor overhaul costs, awaiting management approval	
50006916	Renew B6 20 mm nipple in 34C-406	A34C406	28/09/2009	PZEIDAN	Turnaround - Normal Work		Cannot be done OTR	Yes
50006917	Renew skirt fireproofing on 34C-406	A34C406	28/09/2009	PZEIDAN	Turnaround - Normal Work		Cannot be done OTR	Yes
50006918	Anticipate renewal btm manway on 34C-402	A34C402	28/09/2009	PZEIDAN	Turnaround - Normal Work		Reject, as per P.Zeidan email	No
50006919	Renew skirt fireproofing on 34C-400/401	A34C400	28/09/2009	PZEIDAN	Turnaround - Normal Work		Cannot be done OTR	Yes
50006920	Renew internal trays	A34C402	25/09/2009	LSINCLA	Turnaround - Normal Work		Vessel required to be opened by inspection therefore approve	Yes
50006921	Renew washers on tray hold down bolts	A34C4	25/09/2009	LSINCLA	Turnaround - Normal Work		No other driver to enter vessel therefore reject notification.	No
50006922	Clean Reactor Vapour Overhead Line	A34D4	25/09/2009	LSINCLA	Turnaround - VLF Lines		TSD workscope	Yes
50006923	Replace Flapper Valve Support Bars	A34D4	25/09/2009	LSINCLA	Turnaround - Normal Work		TSD workscope	Yes
50006924	Gritblast and paint fin-fan cowlings	A34E2AX	28/09/2009	MRUTAR	Turnaround - Normal Work		Work cannot be done OTR	Yes
50006925	Gritblast and paint fin-fan cowlings	A34E11AX	28/09/2009	MRUTAR	Turnaround - Normal Work		Work cannot be done OTR	Yes
50006926	Descale and paint lines ex 34E11AX,CX,E	A34E11AX	28/09/2009	MRUTAR	Turnaround - VLF Lines		Work cannot be done OTR	Yes
50006927	Descale and paint shell outlet flanges	A34E2B	28/09/2009	MRUTAR	Turnaround - Normal Work		Work cannot be done OTR	Yes
50006928	Pipe supports are off the O/H piperack	A04L	28/09/2009	MRUTAR	Turnaround - VLF Lines		Work cannot be done OTR	Yes
50006929	Remove, clean and inspect demister pad	A34D7	29/09/2009	LSINCLA	Turnaround - Normal Work		TSD Workscope	Yes
50006930	renew all trays and downcomers 34C-400	A34C400	28/09/2009	PZEIDAN	Turnaround - Normal Work		Damage found in 2006. Supply material for 50% of trays	Yes
50006931	Anticipate renewal of top manway 34C-400	A34C400	28/09/2009	PZEIDAN	Turnaround - Normal Work		Significant damage to both top and middle manway. Constructibility review required	Yes
50006932	Renew N10 & N11 nipples on 34c-400/401	A34C400	28/09/2009	PZEIDAN	Turnaround - Normal Work		Gauging results indicate down to MAT	Yes
50006933	Machine top manway gasket face 34C-414	A34C414	28/09/2009	PZEIDAN	Turnaround - Normal Work		Damage found in 2006. Cannot be done OTR.	Yes
50006934	MPI welds internally on 34C-418	A34C418	28/09/2009	PZEIDAN	Turnaround - Normal Work		Vessel can be done OTR	No

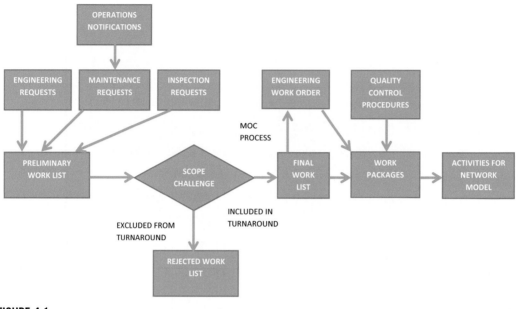

FIGURE 4.1

Work development sequence.

Various scope decision tools are used to assist in the decision-making process. They include the following:

1. **Vibration analysis** equipment and software for pinpointing future repairs/replacements to rotating equipment.
2. **Thickness gauging** equipment and software for calculating **the remaining life** of a pipeline or pressure vessel so that orders can be placed timeously.
3. **Thermovision** cameras for determining the extent of any failure of thermal insulation inside a furnace, lined pipe, or lined vessel so that the most accurate amount of time and resources are planned.
4. Hazard operability studies (HAZOP) and hazard analysis (HAZAN) to determine corrective action that might be undertaken during a turnaround such as a design modification.
5. Root cause analysis (RCA) and fault tree analysis techniques to determine the root cause of particular failures to possibly initiate corrective action during the turnaround.
6. Reliability centered maintenance (RCM) techniques to determine what really needs to be examined during the turnaround.
7. Inspection techniques, such as visual inspection and nondestructive testing (NDT) using ultrasonics, eddy current, etc., to anticipate drafting quality control plans for repair purposes.

The **first three of the tools** above are used extensively to monitor the health of the plant while on stream. Results of these tests should be in the maintenance and inspection records.

The MOC process often encompasses the use of the HAZOP/HAZAN/RCM/RCA items listed above. This process is discussed in *6.8.2 Change control*.

4.2.3 Planning electrical work for a turnaround[9]

As a general rule, the plant's electrical power is the last system to be switched off before turnaround work starts and the first to be switched on once the turnaround has been completed. At times temporary critical power supplies are required and companies who specialize in the supply of temporary power need to be sourced. Early planning to determine power requirements at every stage throughout the turnaround is critical. Long-lead rental items, such as modular diesel generators, also need to be determined.

To avoid unexpected surprises, a checklist of electrical best practices is called for. Some areas on which to focus include safety labeling and personal protective equipment protocols.

Electrical hazards such as arc flashes are a very real concern, which makes proper labeling of electrical equipment and panels an absolute necessity, as well as having a proper lockout procedure in place (Lock Out Tag Out). Arc flash calculations should have been performed at project design stage and displayed on equipment prior to delivery to the job site. Each piece of equipment should be labeled with arc flash decals that clearly show incident energy, flash boundary, and safe working distance values. The arrangement of control panels could pose safety risks. For example, placing high-risk 480 v contactors near any 120 v control circuits that require troubleshooting may result in an incident. Box 4.3 gives an example.

Many of these potential problems could be eliminated by meticulous design, which could have long-term benefits.

> *Example: Schneider Electric promotion.*
>
> *The innovative 2SIS technology in Premset medium voltage switch-gear is ideal for harsh environments and drastically reduces the probability of an internal arc by fully enclosing all current carrying portions of the switch-gear in high dielectric insulating epoxy.*

4.2.4 Anticipating increased work when a vessel is opened

Inputs for review of work requests and assessment methods don't always give the full picture of what to anticipate. Plant upsets since the last turnaround may give further insight as to what to expect. A column disturbance, elevated furnace temperatures, or liquid carryover could be indicators of extra work when opening up a pressure vessel or furnace.

Box 4.3: Serious electrical burns

During a polyester polymer plant turnaround, an electrician was servicing isolated switch-gear on a panel directly above live bus-bars. He dropped a spanner onto the bus-bars, which resulted in a massive arc flash. He was hospitalized with serious burns to the front of his body, and he was blinded by the flash. Luckily his eyesight returned to normal after a few weeks, but he still bears the scars today to remind him of the experience.

Comment: It was not clear as to whether the electrician was wearing cotton or Nomex flame-resistant coveralls. The burns would be less severe with Nomex. Nomex is commonly used offshore and on ships.

Examples

- *Excessive sagging of furnace tubes*
- *Collapse of column trays*
- *Solidification of reactor catalyst.*

Box 4.4 describes the collapse of column trays.

Box 4.5 describes the removal of catalyst after carryover.

4.3 Schedule
4.3.0 Introduction and relevant PM processes

In phase 2 the assembly of the turnaround skeleton model commences. Templates, taken from previous turnarounds, include standard activities. New activities may be added and previous standard activities that are not required may be removed.

The standard processes are:

PMBOK	ISO 21500
6.1b Plan schedule management	—
6.2b Define activities	—

See Appendix A3 for details.

4.3.1 Scheduling

Following from Section 3.3.1, the overall plan is then detailed in a turnaround model.

Box 4.4: Collapse of column trays

Operators noticed a reduction in distillation capacity from a particular column, but it was ignored since the next turnaround was to take place within weeks, and production continued.

When the column was opened in the turnaround, it was found that a number of trays had collapsed. This resulted in major work with removal of the damaged trays, repair and welding in new attachment lugs, and installation of new trays.

Fortunately, the new trays were available at short notice from the supplier who airfreighted them in.

Box 4.5: Liquid carryover

A reactor was taken out of service after the catalyst became inactive. It was discovered that liquid carryover had occurred, which had turned the catalyst to "cement."

Several days were spent jackhammering the catalyst out of the reactor before it could be repaired and recharged.

Minor disruption to production occurred as there were three reactors in the unit with just one in operation at a time while the other two were in various stages of regeneration.

Turnaround model

Project management tools are used throughout the **six** *phase* process. Project management procedures are also developed as progress is made through the phases for the first time, and amended when going through the phases in subsequent turnarounds.

The planning and costing model lies at the heart of the development process and is resident in a LAN-based high-level planning package. This model is built from planning templates of previous unit turnarounds. An outline of this model is shown in Fig. 4.2.

It is essential that all those working on the model use the same standard codes for work breakdown structures, organization breakdown structures, resources, etc. and reference the same standard set of calendars. Each unit "project" can then be "rolled up" into a high-level turnaround "skeleton" project that includes the general sequencing of process unit shutting down and starting up.

The activities in the **pre-turnaround** phase, with required milestones, can also be developed as a separate project and integrated, if or when required, into the model. A summary milestone report can be produced based on the detailed development process at any stage.

Links to other systems could also be useful. The CMMS and related materials management system can provide expenditure commitments when materials are ordered and the time-keeping system can provide labor cost commitments. The accounting system is less helpful, since it generally creates reports only when payments have been made, when it is too late to take corrective action. However, contract control software can also interface with the model for better control of expenditure (see Appendix C12).

With the correct inputs, top management can monitor the turnaround progress with just a few simple reports, including an earned value graph that would show progress and cost to date and the predicted completion and cost. Field supervisors only get what they need to know for the next few days of the turnaround, thereby avoiding information overload at an already stressful time.

The details of critical path method and earned value method (EVM) are discussed in Section 5.3.1, Appendix A4 and Appendix C7.

Computerized maintenance management

CMMSs are essential for effective turnaround management. They generally contain the following modules:

- Work request and work order processing, including work planning and reporting
- Equipment database for historical records of labor and material costs, and material consumption
- Materials stock control
- Materials purchasing.

The use of a CMMS is reported to have the following significant benefits:

- Increase in maintenance productivity
- Increase in equipment utilization
- Decrease in maintenance materials costs
- Reduction in stores inventory

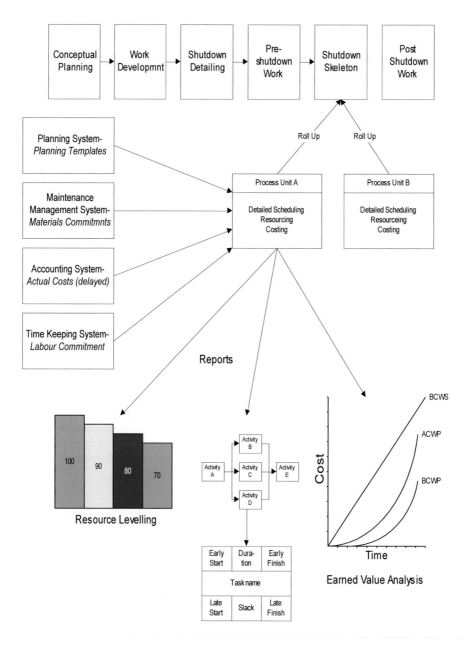

FIGURE 4.2

Turnaround planning model.

- Simplified spare parts management
- Lower total cost of ownership for the asset
- Standardized maintenance best practices.

For turnarounds, the maintenance system offers particularly significant advantages, the main ones being:

1. Templates of the spares required for each item or groups of equipment can be developed for rapid initiation of purchase orders each time a turnaround comes up. Minor editing of a copy of the template is all that is required, thereby reducing manpower significantly once the templates have been established. The standardization of specifications for ordering ensures that one gets exactly what is required. In addition, the moment the order is placed, costs are committed to the turnaround budget thereby enabling the EVM to provide an early and accurate projection of total costs.

2. Historical records of equipment can assist in evaluating the extent of work required for the turnaround and realistic durations can be entered into the critical path network.

4.3.2 Activity definition process

Following on from Section 4.2.1, activities are ready for building into the critical path model. A template from a previous turnaround could be used to build the network diagram (subproject) for a process unit, and only new work needs to be planned from scratch. Unit shutdown and start-up need to be included in the process unit subproject, where overall plant shutdown and start-up would be in the skeleton network for the complex. The complete network is then optimized to come within the expected duration based on previous durations or benchmarking exercises. Once agreement has been reached on the overall duration and resourcing levels, the model is frozen and the baseline is set for progress to be measured against. The process is depicted in Fig. 4.3.

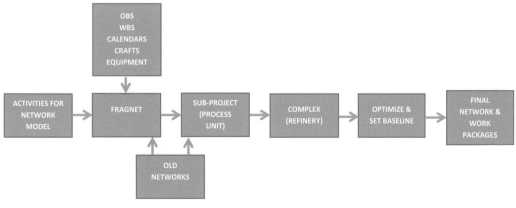

FIGURE 4.3

Activity definition process.

4.3.3 Turnaround activity categorization

Work request deconstruction

Work requests on the final work list are converted to activities for inclusion in a critical path model. Work is categorized before being deconstructed into a network of activities. Categories are as follows:

1. Major tasks
2. Minor tasks
3. Bulk work.

1. Major tasks

These normally require engineering input and are not usually done in every turnaround.

Criteria

- High work content
- Technically complex
- Involves a large number of people with varied skills
- Involves multilevel working
- Take a long time to complete

  Examples

  - *Fluidic catalytic cracking (FCC) cyclone replacement*
  - *Column replacement*
  - *Engineering modifications*

Checklist of requirements

- Major task method statement
- Isolation register
- NDT request and report
- Task safety plan
- Welding procedure
- Pressure test request and report
- Certificate of conformity (boilers, etc.)
- Drawings
- Other documents

Network diagram. Using the above checklist, a network diagram is created for the major task for integration into the process unit network diagram.

2. Minor tasks

Minor tasks require individual specification by an experienced planner and are likely to be repeats of previous turnaround activities.

Examples

- *Exchanger: Pull bundle, replace leaking tubes, assemble and hydro-test exchanger X*
- *Pressure vessel: Inspect and repair pressure vessel Y.*

Checklist

- Small task method sheet
- Isolation register
- NDT request and report
- Welding procedure
- Pressure test request and report.

Network diagram. Using the above checklist, fragnets are created for the minor task for integration into the process unit network diagram. Fragnets are grouped activities for an item of work.

3. Bulk work

Bulk work consists of large groups of identical or similar simple tasks that do not need to be specified in individual task sheets but can be bulked together in one list.

> *Examples*
>
> - *valves*
> - *small pumps*
> - *bursting disks and orifice plates*

Network diagram. These items should be grouped by hydro-test pack where one hydro-test between isolation points can be carried out after completion of all of the items between isolation spades for a common service and pressure rating. Progress is easily monitored.

> *Example*
>
> *Service 10 Low Pressure Steam Valves in Process Unit A.*
>
> *Progress: number of valves serviced to date.*

4.3.4 Planning preparation for access

Access to start physical turnaround work depends on the preparation required for safe access permits to work. The preparation period is the time from stopping production ("feed out") to when the plant internals are isolated, degassed/washed, and decontaminated, ready for human mechanical access. Temperature normalization and decontamination are key elements of the preparation process. Examples of methods used to accelerate this process are discussed below.

> *Example 1*
>
> *Early opening of man-ways on columns, reactors, and furnaces, and the placement of air movers to ventilate the internals (see Appendix C5).*
>
> *Example 2*
>
> *Chemical circulation to passivate internals and/or remove deposits (see Section 4.3.5 item 4).*
>
> *Example 3*
>
> *An LNG plant requires a number of days to warm up from its cryogenic operation temperature to ambient temperature for access. The reverse is true for the start-up process. During these periods*

there is excess flaring and various operational techniques are applied to reduce this time and consequent flaring.

The planning and sequencing of the process for shutting down and preparation for access is crucial to keeping the overall turnaround duration to a minimum and should be included in the skeleton network. An example is shown in Box 6.3. Preparation for access is discussed further in Section 7.7.2.

4.3.5 Innovative ideas for reduction in turnaround duration

Once the scope of the turnaround has been agreed, the turnaround duration has to be reviewed and reduced as much as can be safely achieved. Innovative solutions are sometimes needed when the critical path turns out to be excessively long. Some techniques, tools, and methods that can reduce activity time considerably are:

1. FCCU riser internals
1a. Inspection: abseiling
In a particular case, the inspection duration was considered to be too long due to the time required to install scaffolding. The turnaround planning engineer came up with the idea of using abseiling techniques, thereby avoiding the use of scaffolding altogether. The inspectors were trained in the use of rope techniques and the inspection duration was drastically reduced.

1b. Refractory removal
The removal of refractory from inside the FCCU riser can be speeded up with the use of a remote controlled hydro-blaster.

2. Spading and de-spading of flanges
2a. Flange spreading tool
A flange spreading tool[2] is used for the spreading of flanges for spading, and aligning flanges for bolt up. Spading and de-spading time is considerably reduced using this tool.

2b. Tag/spade swapping
Prior to a turnaround, operations and maintenance staff identify various spading points for the isolation of sections of process pipe and plant in preparation for opening. A tag is placed at each location, identifying what spade, and related pressure rating, is required. At the time of the turnaround, the process of swapping a tag for a spade is used, and all tags/spades are displayed on a tag/spade board, which reduces the risk of leaving a spade in the plant during start-up. When the spaded line is safe to work on, the board displays only tags, but at start-up, the board should display only spades (see Section 7.5.5).

3. Heat exchanger bundle extraction, cleaning, and hydro-testing
3a. Heat exchanger bundle puller
The use of a heat exchanger bundle puller cradle, for extracting large tube bundles, speeds up the heat exchanger cleaning and hydro-testing cycle. However, the use of just a single cradle could cause a bottleneck when all the bundles are installed at the same time toward the end of the turnaround.

3b. Very high pressure cleaning equipment

Very high pressure (as opposed to just high pressure) water cleaning speeds up the bundle cleaning process. Adequate numbers of these very high pressure machines are required to suit the number of exchangers that need to be cleaned.

3c. Removal of complete exchanger

The removal of complete heat exchangers to a workshop environment where a production line operation can be set up could improve productivity considerably.

> *Example of removing the complete heat exchanger from the process area.*
>
> *Bahrain Refinery has the practice of removing entire heat exchangers to their workshop for cleaning and hydro-testing, thus decongesting the area of work in the process plant, and increasing productivity by creating a mini production line in the workshop. The refinery has, of course, been designed for this practice.*

4. Cleaning of hard-to-remove deposits
4a. Chemical cleaning

Chemical cleaning could be applied to "hard-to-remove" deposits. The cleaning cycle of spading, pump-around, passivating, and de-spading needs to be built into the preparation phase of the turnaround. Only those professional cleaning companies who have successfully completed similar work before should be considered. Box 4.6 gives an example of chemical cleaning gone wrong.

5. Modular replacement
5a. Mechanical seals

On a small scale, mechanical seals can be removed and replaced with reconditioned seals with the proviso that the removed seals are sent to an original equipment manufacturer–qualified seal repair contractor for reconditioning between turnarounds.

5b. Gas turbines

On a much larger scale, a gas turbine could be swapped out for an identical rebuilt turbine. The turbine that has been removed is then sent to the vendor for rebuilding for the next turnaround.

Box 4.6: Chemical cleaning gone wrong—polyethylene terephthalate (PET) plant

During a turnaround, a company employed a contractor to chemically clean the nitrogen plant air compressor heat exchangers. (Nitrogen is required to flush the process piping at start-up and blanket the process vessels during operation.)

The cleaning contractor performed the task, but, on start-up, it was found that the compressors could not produce air for the nitrogen plant and so the start-up was delayed.

Pinhole leaks in the air compressor heat exchangers, caused by incorrect cleaning methods, were discovered. This resulted in all air compressor heat exchangers having to be replaced. Luckily the compressor manufacturer (Atlas Copco) had sufficient number of spares, but more than a week's production was lost due to this incident.

Lessons learned
1. Ensure the contractor is qualified to perform this specialized sort of chemical cleaning
2. Visit sites where the contractor has done similar work to be assured of their capabilities, before awarding the contract
3. Request and verify their process for cleaning and monitor their progress to ensure that they work to the process
4. Ensure that the contractor's appointed supervisor has successfully performed many similar cleaning operations before.

Example

Qatargas has a maintenance agreement with Al Shaheen GE Services Company for the exchange of gas turbines.

5c. Tube bundles
The exchange of tube bundles for criticality A exchangers is essential, especially if these are the cause of an unplanned shutdown.

5d. Relief valves
If this practice of replacement is adopted, it is essential to retain the history of each individual valve to be able to identify each one in the inspection records.

6. Heavy lift alternatives
Some heavy lift alternatives are described in the Refs.[3,4].

7. New radiography method: Saferad[5]
This method allows both radiographic testing (RT) and mechanical work to continue as the barriers are down to 1 m from source (50 m for traditional RT). Savings include earlier start-up of plant as no night shift is required for radiography.

Example[6]

During the last Abu Dhabi National Oil Company (ADNOC) Liquefied Natural Gas (LNG) turn-around, 700 joints were RTs using Saferad, resulting in major cost savings.

8. Flare inspection using a drone (UAV)[6]
Drones are used to undertake visual inspection of items such as flare tips. Time and cost savings are significant: scaffolding, crane, man-cage, etc.

9. Corrosion under insulation inspection using OpenVision[7]
Inspection of corrosion under insulation traditionally required the removal of the insulation. OpenVision is a lightweight live video X-ray imaging system specifically designed for handheld inspection. This portable system is now commonly used for corrosion under insulation inspection.

4.4 Cost
4.4.0 Introduction and relevant PM processes
In phase 2, the cost estimate is more accurate since the turnaround scope has been better defined.

Relevant standard processes are:

PMBOK	ISO 21500
7.1b Plan cost management	—
7.2b Estimate costs	4.3.25 Estimate costs
7.3b Determine budget (prelim)	4.3.26 Develop budget

See Appendix A4 for details.

4.4.1 Cost development

An intermediate cost estimate may be needed if an amendment to the annual budget is required. This occurs when the scope has grown to a point where a contingency is not sufficient to cover additional work costs. It is for this reason that the scope challenge process is so important to avoid any unpleasant surprises. The estimate is based on the preliminary critical path model for the upcoming turnaround and thereafter requires a **preliminary budget decision**.

At this stage, factoring the man-hours cost to obtain total cost may no longer be appropriate as more detailed man-hours are now available and separate materials, tools, equipment, and overheads costs can be calculated.

Note that everything related to the turnaround should be included in the cost estimate, even operator costs. Operators are often redeployed as hole watchers, etc. for the duration of the turnaround, thus their time can be charged to the turnaround.

4.5 Quality

4.5.0 Introduction and relevant PM processes

Quality management planning continues in phase 2. Quality records from both the previous turnaround, and onstream records, are extracted for review and input into the scope challenge process.

The relevant project standard processes are:

PMBOK	ISO 21500
8.1b Plan quality management	4.3.33 Plan quality

See Appendix A5 for details.

4.5.1 Quality development

Generic quality control plans are updated or written.

Basic quality requirements are outlined in Fig. 4.4.

4.5.2 Quality records

Quality records from the previous turnaround as well as onstream inspection records are extracted for the turnaround scope review process. These records include:

- Corrosion current graphs
- Remaining thickness calculations
- Valves' actual conditions (if passing, etc.)

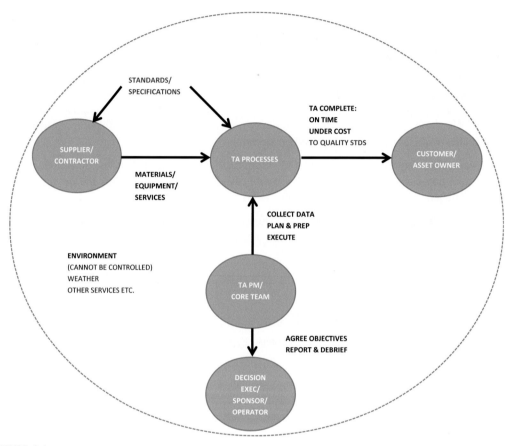

FIGURE 4.4

Basic quality requirements.

- Status of flame, gas, smoke detectors
- Annual preventive maintenance records for distributed control system, fire and gas systems, emergency shutdown systems.

4.6 Resources
4.6.0 Introduction and relevant PM processes

In phase 2 early resource leveling, using a template from the previous turnaround is required to estimate man-hours for each trade so as to give potential contractors some indication of what will be required. Follow-up on the ordering of critical specialist services and equipment and long-delivery items is also required.

Related standard processes are:

PMBOK	ISO 21500
9.2a Estimate activity resources	4.3.16 Estimate resources
9.3b Acquire resources (project team)	–
9.4a Develop team (core)	4.3.18 Develop project team

See Appendix A6 for details.

4.6.1 Resource leveling 1

The critical path model used in previous turnarounds on the same plant is used to generate generic resource requirements.

> *Example*
>
> *Estimate of total man-hours for welders required and the maximum number of welders required on-site at the same time.*

Selection of the core team is completed based on the requirements listed in Section 3.6 and with a view to ensuring that this core team demonstrates comprehensive complementary skills.

4.6.2 Early ordering activities

Critical equipment, tools, and services

Further to Section 3.6.8, firm orders need to be placed for required equipment and tools.

Long-lead items

Further to Section 3.9.2, long-lead orders need to be tracked to ensure timely delivery.

4.6.3 Planning for critical resources

Early identification and timely procurement of critical and scarce resources is required. Specialized services were discussed in Section 3.6.8. Potential shortages of critical crafts need to be identified. Crafts include:

- Instrument technicians (for calibration, trip and alarm testing)
- Medium voltage (3.3–11 kV) electricians
- Protective coating specialists
- Specialty and alloy welders (see Section 5.6.3 *example*)
- Specialized millwrights and boilermakers
- Inspectors.

Actions to be initiated include:

1. Identifying "order of magnitude" requirements for critical and scarce resources
2. Ensuring multiple contractors for providing critical resources

3. Obtaining scarce resources from affiliate companies (see Section 6.5.2).
4. Early involvement of contractors in developing plans to overcome bottlenecks due to craft shortages
5. Procure critical resources from specialist contractors
   a) Instrument technicians from an instrument contractor/instrument calibration authority
   b) Licensed high-voltage electricians from licensed high-voltage contractors
   c) Protective coating specialist from specialist coating contractors
   d) Welders from a tank/pressure vessel manufacturer
   e) Boilermakers and millwrights from a steel fabricator/pressure vessel manufacturer
   f) Inspectors from a registered inspection authority.

4.7 Communication

4.7.0 Introduction and relevant PM processes

In phase 2, the logistics for the turnaround are established. Work list, major task review, inspection review, and project work review meetings are initiated.

The relevant standard processes are:

PMBOK	ISO 21500
10.1b Plan communications management	4.3.38 Plan communications

See Appendix A7 for details.

4.7.1 Logistics

Maintenance department usually coordinates the use of available plot space for cranes, trailers, prefab areas, toilets, change rooms, etc. It is essential to document who is responsible for which areas, who has access to these areas, the exact locations of portable accommodation, stores, etc. and at what times access is available. A marked-up plot plan is useful for display and approval purposes (see Section 5.7.3).

Some of the more important items to note are:

1. Streamlined system of security clearance for the massive influx of construction workers.
2. Approval of contractor staff qualifications, especially welders, needs to be done well in advance of the turnaround.
3. Catering and first aid for the masses, available 24 h a day.
4. Radios and allocated radio frequencies.
5. Accommodation for change rooms, washrooms, toilets, mobile offices, and stores (materials and tools) close to the action.
6. Mobile equipment such as cranes, compressors, welding machines, hydro-blasting machines, sandblasting machines, breeze generators, waste skips, heat treatment rigs, vacuum tankers, diesel tankers, etc.

7. Services such as waste disposal, lighting and power, water (including drinking water), steam, air, phone lines, etc.
8. Safety equipment such as breathing apparatus, safety signs, harnesses, ropes, etc.
9. Training of fire and hole watchers, permit issuers, permit receivers, etc.
10. Staging list and scaffolding storage close to the action.
11. Consumables such as clean sand, sandbags, grit, barrier tape, drums/bags for waste, rods and gases for welding and cutting, dry ice, diesel for powering mobile equipment, copperslip for threads, etc.
12. Air, steam, and water hoses, road plates, etc.
13. Special heavy lift cranes booked and locations marked out.
14. Documents—stationery, change order forms, shift turnover books, hydro-test record forms, copies of specifications and drawings, etc.
15. Electronic devices—tablets, etc.
16. Blinds, blind lists, tagging for blinding, and blind racks.
17. Special materials such as exchanger and man-way gaskets and bolts, de-mister pads, trays, etc.
18. Special tools such as tube rollers, reamers, hydro-test rings, portable valve seat cutters, etc.
19. Rigging gear such as slings, winches, chains, lifting frames, etc.
20. Catalysts—dispersion rigs, replacement methods, and storage, handling, and disposal of spent catalyst.

Box 6.1 shows Bapco's T&I checklist.

4.7.2 Review meetings
Meetings related to scope review are initiated (see Box 3.14 for list of meetings).

4.7.3 Temporary power[9]
Typically the supplier of temporary electrical power conducts a walk-through of the plant with the plant's engineers. The walk-through is designed to give the vendor a detailed understanding of which plant needs to be kept running in the turnaround, how much load is required, siting of the temporary generators, switch-gear and cable runs, and the best fueling options. With this information the vendor can then design and specify the most efficient temporary power system for that specific turnaround. One-line diagrams help identify locations of generators, switch-gear, and cabling. Easy access to generators for refueling and servicing needs consideration.

4.8 Risk management
4.8.0 Introduction and relevant PM processes
In phase 2 the risk management process for the forthcoming turnaround is initiated. This is based on ISO 31000 Risk management.

An audit at the end of this phase determines that preparation in line with the agreed milestone schedule is on target.

The related standard processes are:

PMBOK	ISO 21500	ISO 31000 risk management
11.1a Plan risk management	–	1. Establish the context
11.2 Identify risks	4.3.28 Identify risks	2. Communicate and consult
11,3 Perform qualitative risk analysis	4.3.29 Assess risks	3. Risk assessment
11.4 Perform quantitative risk analysis		Risk identification
		Risk analysis
		Risk evaluation
11.5 Plan risk response	4.3.30 Treat risks	4. Risk treatment
11.6 Implement risk responses		

See Appendix A8 for details.

4.8.1 Purpose of risk management

The purpose of turnaround risk management is to:

1. Specifically *identify* factors that are likely to impact the turnaround objectives of scope, quality, time, and cost
2. *Quantify* the likely impact of each factor
3. Give a *baseline* for turnaround noncontrollables
4. *Mitigate* (make less severe) impacts by exercising influence over project controllables.

Example

*The weather is a major risk factor for open plant turnarounds (**identification**). Rain could delay work for days. After reviewing weather data (**quantification**), a suitable time of the year is determined as the best time to do the turnaround. This is setting a **baseline** for noncontrollables. A number of tarpaulins are then retained in stock in case there is heavy rain so that certain critical items can be covered (**mitigation**).*

4.8.2 Project risk definition

Project risk is the cumulative effect of the chances of uncertain occurrences adversely affecting project objectives.

The risk function can be broken down into four processes:

1. Identification
2. Analysis and evaluation
3. Response/treatment
4. Documentation

1. Identification

The various tools discussed in Appendix A8 assist in identifying items at various levels of risk. Brainstorming could be used to identify all those possible risks that may significantly impact the success of the turnaround.

Examples of higher risk profiles

- *Areas with a high manpower concentration of people working in very close proximity to each other have a high safety risk.*
- *A certain amount of refractory is planned for replacement in a furnace based on thermal photographs and historical records, but when the furnace is opened much more work is required than planned for.*

2. Analysis and evaluation (assessment)

Analysis and evaluation of risk can be used as a team-building exercise.

All project risks are characterized by the following *three* risk factors:

1. Risk event—precisely what might happen to the detriment of the turnaround.
2. Risk probability—how likely the event is to occur.
3. Amount at stake—the severity of the consequences.

With this data a ranking can be given for a particular risk as follows:

$$\text{Risk Event Ranking} = \text{Risk probability} \times \text{Amount at stake}$$

When activities are placed into the critical path network, their risk can be more clearly assessed. Activities on or near the critical path need to be scrutinized more carefully as to the probability of extending the duration of the turnaround and/or causing a safety/health/environmental/quality (SHEQ) incident. Details of calculating cost contingency based on risk are discussed under Contingency in Section 5.4.3.

An example of a risk assessment matrix is shown in Box 4.7.

3. Response/treatment

Once the risks have been assessed, alternative responses for the high-risk items need to be reviewed.

Examples

- *A contract with a specialist contractor to carry out FCCU slide value maintenance during a turnaround rather than doing the work internally or with a general contractor. This is known as risk deflection. (Comment: Risk deflection must improve the probability of achieving the required turnaround duration, or even shortening it, to be of value.)*
- *Chemical cleaning in preparation for vessel entry is a high-risk activity that speeds up the cleaning process prior to obtaining access. (For example, the vessel could be pitted by the chemicals or operators could be chemically burned.) Ensure there are procedures in place that must be strictly adhered to and ensure staff are suitably trained. Alternatively, conventional cleaning may be a better alternative if the relevant vessel is not on or near the critical path.*

The distribution of contractual risks is discussed in Sections 3.9.4 and 3.9.5.

Risk analysis and evaluation (assessment) and response/treatment during scope challenge workshops are outlined in Box 5.4.

Box 4.7: Risk assessment matrix (RAM) example[1]

LIKELIHOOD	5 MINOR (<S10K)	4 LOW (S10K-100K)	3 MEDIUM (S100K-500K) CONSEQUENCE	2 MAJOR (S500K-1M)	1 CATASTROPHIC (>S1M)
1 HIGH (>75%)	3	3	1	1	1
2 SIGNIFICANT (50-75%)	3	3	1	1	1
3 MEDIUM (25-50%)	6	5	4	2	2
4 LOW (10-25%)	6	6	5	5	4
5 NEGLIGIBLE (<10%)	6	6	6	5	5

note: costs include LPO

1	Tier 1	Risks that are very likely to happen, with High Consequences
2	Tier 2	Significant risks to the shutdown, with high consequences
3	Tier 3	Mid level risk with high likelihoods
4	Tier 4	Mid level risk with high consequences
5	Tier 5	Low risks
6	Tier 6	Risks that have low impact on the shutdown & Refinery and don't require mitigation

Tier 1

Write and release work order stating that it is pending inspection (work does not start until given direction).
Indude detailed job steps for this work in the Primavera Plan
Resources are allocated in the plan and will be available
Constructability/Schedule Optimisation carried out on mitigation plan for ones with consequence 1 & 2
Purchase required materials and carry out any pre-fabrication if deemed prudent
Put a very high priority on opening and inspecting to allow for quick discovery
Indude in final estimate
Advise refinery management of the ones with consequence 1 and 2 and of their mitigations, especially if high expenses are occurred to mitigate them.
WMS approval by superintendents and refinery manager

Tier 2

Write and release work order stating that it is pending inspection (work does not start until given direction).
Materials are sourced or on consignment or "on hold" in supplier's stores. Possibly buy if risk is considered too great
Put a high priority on opening and inspecting to allow for quick discovery
Include detailed job steps for this work in the Primavera Plan
Contact required contractors in advance – confirm availability

4. Documentation

It is essential to document high-risk items and the actions to be taken in the event of an incident, so that quick action is assured during the turnaround. Support for the cost and schedule contingencies that have been catered for in the budget and schedule is also required (see Sections 5.3.2 and 5.4.3). This documentation should be included in the final turnaround report in order to reduce risk management work for the next turnaround.

Risk is discussed in more detail in Section 5.8.2 where establishment of a formal risk register is described. Also Section 5.4.3 discusses cost contingency using a risk assessment technique.

It is always preferable to spend considerably more time during the planning stage on risk management when the amount at stake is minimal, as this amount could escalate substantially during the actual execution phase.

4.8.3 Hazard management

As the scope evolves, related hazards become more apparent. It is recommended that these be listed and relevant management approaches identified. An example is shown in Box 4.8. They also need to be linked to the risk register (see Section 5.8.2).

4.8.4 Audit

As discussed in Section 3.8.8, an audit is recommended at the end of phase 2.

4.9 Procurement
4.9.0 Introduction and relevant PM processes

Constructability and alternative work methods need to be studied as work develops in phase 2, so as to minimize the turnaround duration.

Orders for all long-delivery items are placed at this stage and closely tracked to ensure delivery before the start of turnaround execution.

Relevant standard processes are:

PMBOK	ISO 21500
12.1a Plan procurement management	4.3.35 Plan procurements
12.2a Conduct procurements	4.3.36 Select suppliers

See Appendix A9 for details.

4.9.1 Procurement planning

A comprehensive procurement plan is established for services and materials, detailing contract bundles for tendering and listed materials (grouped). Potential services contractors and materials vendors are identified. The plan should also identify who is going to order what and when: engineering, maintenance, operations, or materials departments.

Example

1. *Engineering orders new furnace tubes, columns, etc.*
2. *Maintenance orders pump bearings, mechanical seals, valves, etc.*
3. *Operations orders catalysts, chemicals, etc.*
4. *Materials arranges stock items (with increased quantities for the turnaround).*

The logistics coordinator, as part of the core project team, needs to establish all hire contracts: cranes, scaffolding, food services, temporary buildings, temporary power, mobile equipment (welding machines, air compressors, air movers, electricity generators, fuel tenders), cars, trucks, etc.

Box 4.8: Hazard management example[1]

CAL TEX T&I SHUTDOWN CDU 2010

SYSTEM	NO.	HAZARDS							DECONTAMINATION								HAZARD MANAGEMENT			HANDOVER					HAZARD MANAGEMENT AFTER HANDOVER							
		H2S	LEL	Sour Water	H/C	Pyrophoric	Mercury N2	Other	Steam Out Cold Circ	Water Steam Out Wash	H2 Purge	N2 Purge Interm Random Purged	Na2CO3	Vent	Drain	Air Purge	Other	Fresh Air to blind	Half face resp (Hg) during steam out	H2S Monitor	Gas test	Sample & test if reqd.	O2 20.9%	LEL Nil	H2S Nil	Conf low point drain	Other	Fresh Air	Half face Resp Air (Hg)	H2S Monitor	Gas Test	Sample & test if reqd.	Manuf review of VMS	Verify equip. depress and drained	Water on demister pad	Restricted entry to hazardous area eg reaction pad O2 N2
Crude Feed U/L to Desalter C-113 7/S E-101, 7/S E-103A, 7/S E102A, 7/S E102B, 7/S E103A, 7/S E105	3a / b	L	H	L	H	L	L		Y	Y	N	N	N	Y	Y	N		N	Y	N	Y	Y	Y	Y	Y	Y		N	N	N	N	Y	Y	Y	N	N
Desalters: 45C-113 Primary Desalter	3c	L	H	L	H	L	L	SULPHUR/ ELECTRICAL	Y	Y	N	N	N	Y	Y	N		N	Y	N	Y	Y	Y	Y	Y	Y		Y	N	N	N	Y	Y	Y	N	N
45C-120 Secondary Desalter	3c	L	H	L	H	L	L	SULPHUR/ ELECTRICAL	Y	Y	N	N	N	Y	Y	N		N	Y	N	Y	Y	Y	Y	Y	Y		Y	N	N	N	Y	Y	Y	N	N
Crude Feed From 45C-120 to C-100 7/S E104, 7/S E106, 7/S E134A, 7/S E124B, F100X,	3d	L	H	L	H	M	L		Y	Y	N	N	N	Y	Y	N		N	Y	N	Y	Y	Y	Y	Y	Y		N	N	N	N	Y	Y	Y	N	N
Atmos Column C-100 C-100,S/S E-124B, S/S E-134A,S/S E-103B, E-122S	4a	L	H	H	H	M	L	ASBESTOS*	Y	Y	N	N	N	Y	Y	N		N	Y	N	Y	Y	Y	Y	Y	Y		N	N	N	N	Y	Y	Y	N	N
C-100 Reflux and Product Rundown C-117, C-116, C-115	4b	L	H	H	H	M	L		Y	Y	N	N	N	Y	Y	N		N	Y	N	Y	Y	Y	Y	Y	Y		N	N	N	N	Y	Y	Y	N	N
Line No.3 Sidestream: 7/S E-103, S/S E-106, S/S E-131, E-108*S	4a	L	H	H	H	M	L	ASBESTOS*	Y	Y	N	N	N	Y	Y	N		N	Y	N	Y	Y	Y	Y	Y	Y		N	N	N	N	Y	Y	Y	N	N
Line BCR: 7/S E-112, S/S E-105, S/S E105A	4b	L	H	H	H	M	L		Y	Y	N	N	N	Y	Y	N		N	Y	N	Y	Y	Y	Y	Y	Y		N	N	N	N	Y	Y	Y	N	N
Line No.2 Sidestream: C-102, S/S E-104, S/S E/101, C-107	4a	L	H	H	H	M	L	ASBESTOS*	Y	Y	N	N	N	Y	Y	N		N	Y	N	Y	Y	Y	Y	Y	Y		N	N	N	N	Y	Y	Y	N	N
Line No.1 Sidestream: C-101, E-110AB	4b	L	H	L	H	M	L		Y	Y	N	N	N	Y	Y	N		N	Y	N	Y	Y	Y	Y	Y	Y		N	N	N	N	Y	Y	Y	N	N
Line MCR 7/S E-133,S/S E-103A	4b	L	H	H	H	M	M		Y	Y	N	N	N	Y	Y	N		N	Y	N	Y	Y	Y	Y	Y	Y		N	N	N	N	Y	Y	Y	N	N
Line TCR: S/S E-102A, S/S 102B	4b	L	H	H	H	M	M		Y	Y	N	N	N	Y	Y	N		N	Y	N	Y	Y	Y	Y	Y	Y		N	N	N	N	Y	Y	Y	N	N
Line Atmos. O'heads to 45C-104: E-100*S	5a	L	H	L	H	M	L		Y	Y	N	N	N	Y	Y	N		N	Y	N	Y	Y	Y	Y	Y	Y		N	N	N	N	Y	Y	Y	N	N
Vessel C-104, C-105 and line to 45C109: 7/S E111B, 7/S E-111A, 7/S E131	5b	L	H	L	H	L	H		Y	Y	N	N	N	Y	Y	N		N	Y	N	Y	Y	Y	Y	Y	Y		N	N	N	N	Y	Y	Y	N	N
Benzin 45C-109:	5c	L	H	L	H	M	H	ASBESTOS*	Y	Y	N	N	N	Y	Y	N		N	Y	N	Y	Y	Y	Y	Y	Y		N	N	N	N	Y	Y	Y	N	N

4.9.2 Constructability and alternative work methods

Definition of constructability[8]

The effective and timely integration of construction knowledge into the conceptual planning, design, construction, and field operations of a project to achieve the overall project objectives in the best possible time and accuracy at the most cost-effective levels.

All interested parties need to confer to find the most effective solutions to the more complex activities that are to be undertaken. An on-site investigation is essential and communication with maintenance personnel is crucial. Experienced maintenance personnel can advise on a number of pragmatic lessons learned from previous exposure that can be of benefit.

Some items to note when determining the best options:

1. Standardization of work
2. Prefabrication, preferably in a workshop or factory environment (see Box 6.10, for example)
3. Easy accessibility to work areas
4. Construction techniques that facilitate efficiency and speed (see Box 4.9)
5. Optimization of work sequences
6. Use of standardized similar equipment (example: a standard 20-ton mobile crane).

According to the critical path analysis, various activities might need to be done simultaneously in the same area. A reevaluation of the sequence of activities or an assessment of the work actually being undertaken could help to reduce the activity intensity in the area. Common resources such as cranage might also be better utilized for the benefit of all the activities in the area.

> *Examples of alternative methods used to speed up the turnaround:*
>
> * *The removal of the top of a FCC reactor with attached cyclones using a hydro-abrasive cutting tool that cuts through the steel and refractory, and the installation of a complete replacement that has been prefabricated prior to the turnaround, has been found to be far quicker than removing and installing each cyclone through a small man-way with associated removal and replacement of refractory.*
> * *The chemical decontamination of process units to obtain access earlier than when using conventional methods such as flushing with diesel, water, etc.*

Box 4.9 describes two alternative methods used for FCC cyclone replacement. A third method is described in Box 6.6.

Other ideas are listed in Section 4.3.5. The associated risks also need to be evaluated for these alternative work methods as per Section 4.8.

4.9.3 Long-delivery items

Provisional orders will have been placed in phase 1 for long-delivery items (see Section 3.9.3) and these orders now need to be confirmed.

Box 4.9: Alternative methods for FCCU cyclone replacement

Method 1: Bahrain Petroleum Company (Bapco)—side entry

Bapco has the oldest fluidic catalytic cracking unit in the world. It was built in 1945 to increase production for the war effort in World War II.

In the 1990s, replacement of the cyclones was planned and Foster Wheeler was contracted to assist with developing the best method for minimal downtime during the planned turnaround.

Various approaches were discussed.

The "traditional" method via the existing main vessel man-way was ruled out as it took too long and extended the duration of the turnaround excessively.

Removal of the top of the reactor with high-pressure sand/water hydro-blasting was also ruled out as this reactor has a very large diameter and there wasn't a big enough crane in the region to lift the top with the old cyclones attached.

The solution decided upon was to replace the cyclones through a large side man-way, which Bapco estimated would save up to 18 days.

In preparation, a platform was built next to the anticipated location of the large man-way while the plant was still online, and a refractory lined wall plate with a man-way was prefabricated for fitting during the turnaround. In the turnaround, the wall was cut out and the wall plate and large ready-made man-way were installed and the cyclones were then replaced through this man-way. FCCU turnaround duration was reduced considerably as this work was carried out in parallel to the required routine maintenance and inspection work.

The photo and illustration show the wall plate being lifted into place and the lifting and cranage strategy for installing the cyclones.

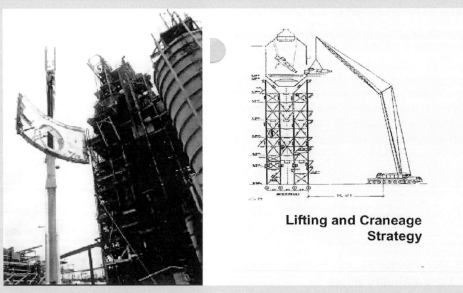

Lifting and Craneage Strategy

Method 2: Qatar Petroleum (QP)—top removal

Damaged cyclones were due to be replaced in Qatar Petroleum's refinery during the 2005 turnaround. The methodology of cutting the top off the reactor using high-pressure sand/water hydro-cutting had been well established by then, so QP chose this method.

A top specialist FCC cyclone manufacturer was awarded the contract to manufacture a new reactor top with the cyclones attached, and a contractor with extensive FCC cyclone replacement experience was subsequently selected to install them.

Box 4.9: Alternative methods for FCCU cyclone replacement—cont'd

The duration of the installation was planned using PrimaVera and honed to 30 days. At that time, the contractor was undertaking a number of FCC cyclone replacement projects around the world and the typical time taken for the work was 30 days. It was to be carried out in parallel with the normal turnaround maintenance and inspection work, thus not necessarily increasing the total duration of the turnaround.

The lift is shown in the photo below.

The replacement was successfully carried out in the agreed time frame, but the total duration of the turnaround was deemed to be excessive. This is discussed separately in Box 5.6.

Example

Furnace tubes and alloy process piping are normally long-delivery items.

Long-delivery items, as well as other critical purchases, need to be tracked from bid to receipt on site. Box 3.16 gives an example of a long-delivery item having to be airfreighted as approval to purchase was not received in time for sea shipment.

4.9.4 Materials contingency planning

There are a number of materials factors that can adversely affect the plan. These include:

1. Materials not available from a known source
2. The required quantity of material cannot be sourced from a single supplier
3. Delivered materials may not meet specification
4. Materials can deteriorate or be destroyed while stored on site.

Box 4.10 gives an example of storage of sensitive electronic equipment on-site prior to installation in a control room.

Since it is not possible to plan for every eventuality, a contingency plan is required for materials. Contingency planning comprises two parts:

1) Quantity contingency, and
2) Availability contingency.

Quantity contingency makes allowance over the specified requirements for wastage, spoilage, etc.

Availability contingency ensures that those materials that are difficult to obtain and/or are required at short notice are delivered to site to be available when required.

An emergency procurement plan would be useful for key items that could be required at short notice.

4.10 Stakeholder management

4.10.0 Introduction and relevant PM processes

At the end of phase 2, the decision executive is required to approve the preliminary budget if costs and schedule are significantly different from that already approved as part of the business plan and budget.

Other stakeholders need to be notified of the timing as well as the expected total manpower requirements (see also Section 3.6.7).

Related standard processes are:

PMBOK	ISO 21500
13.3 Manage stakeholder engagement	4.3.10 Manage stakeholders

See Appendix A10 for details.

Box 4.10: On-site storage to prevent deterioration of electronic equipment

The Bahrain Petroleum Company (Bapco) was upgrading the distributed control system of a refinery complex, but the project was running late and the new control room was not ready to receive the vast amount of delicate electronic equipment that had already been delivered to site.

No contingency plans for storage had been made and so it was opportune that a new cafeteria had just been built and the air-conditioned old cafeteria was available as a store.

4.10.1 Decision executive

The preliminary budget decision by the decision executive includes:

1. The decision executive being appraised of the duration and probability of meeting the target. This relates directly to production capability and market need.
2. Approval of the ±20% budget.
3. Endorsement of the scope, as per the updated preliminary work list.
4. Risks that need to be addressed.

If the initial budget proposal is sufficiently accurate, further formal approval may not be required at this stage.

4.10.2 Other stakeholders

Other stakeholders, including contractors and other industries in the area, need to be notified of the timing and duration of the turnaround, and the amount of resources necessary. It is not prudent to have a number of large turnarounds on the go in the same area simultaneously. Thus industry in the area or country tends to collaborate with respect to the timing of their respective turnarounds (see also *1.6.3 Other definitions − Portfolio management example*).

Raw materials suppliers as well as customers also need to be notified of the production stoppage.

4.11 Summary: phase 2—turnaround work development
4.11.1 Focus areas

1. Firm up all turnaround inputs—capital jobs, HAZOP studies, root cause analysis, maintenance requirements, legal requirements, SHEQ requirements, operator suggestions, process engineering requirements, etc.
2. Conflict resolution
3. Constructability input

4.11.2 Deliverable

Requirements for the **PBD** include:

- Preliminary critical path schedule
- Turnaround duration
- ±20% cost estimate
- Updated preliminary work list
- Other items of interest, including risk issues.

4.11.3 Tools

1. Conflict resolution
2. Critical path scheduling
3. Auditing

4.11.4 Timing

6−12 months prior to turnaround.

References

1. McGrath R. *The recipe for success: investing in comprehensive shutdown planning.* STO Asia 2014 IQPC.
2. Equalizer International Group. Available from: www.equalizerinternational.com.
3. Mammoed crane lift example. Available from: http://www.hydrocarbonprocessing.com/news/2018/05/reducing-downtime-by-building-up-high.
4. Mammoet heavy lift example. Available from: http://www.hydrocarbonprocessing.com/news/2018/05/mammoet-replaces-400-ton-coke-drums-at-germanys-largest-refinery.
5. Saferad. Available from: http://www.saferad.com/.
6. ADNOC presentation: optimizing inspection activities during periodic shutdowns to avoid delays STO MENA. Abu Dhabi: UAE; 2017.
7. OpenVision. Available from: https://qsa-global.com/product/openvision-ovcf-ndt/.
8. Construction Industry Institute. Available from: www.construction-institute.org/resources/knowledgebase/best-practices/constructability.
9. *What you need to know when installing critical power utilities during a plant turnaround.* Hydrocarbon Processing; December 2017.

Phase 3 Detailed planning

5.0 Outline

This phase starts with the approved preliminary budget. The critical path model is built using inputs from the agreed scope of work. Details of how the model is built and what to expect from it are discussed and agreed by interested parties.

The gate at the end of this phase is normally referred to as the **final budget decision (FBD)** where the scope is totally integrated into the critical path model so as to determine the exact duration and cost of the turnaround.

This detailed planning phase (phase 3) pulls all the details together for approval by the decision executive, generally in the form of an updated charter.

Subjects covered are:

1. integration
2. scope
3. schedule
4. cost
5. quality
6. resources
7. communication
8. risk (including health, safety, and environmental issues)
9. procurement
10. stakeholder management

Specifically, Table 5.1 references PMBOK and ISO 21500 processes applicable in this phase.

5.1 Integration

5.1.0 Introduction and relevant project management (PM) processes

Phase 3 entails detailing the project management (PM) plan down to the nearest hour for execution. Details for integrating capital projects are also reviewed and planned.

Turnaround Management for the Oil, Gas, and Process Industries. https://doi.org/10.1016/B978-0-12-817454-8.00005-8

Table 5.1 PMBOK and ISO 20500 processes.

No	PMBOK	ISO 21500
1	4.2c Develop project management plan	4.3.3 Develop project plans
2	5.4 Create WBS	4.3.12 Create WBS
	5.5a Validate scope	4.3.13 Define activities
3	6.3 Sequence activities	4.3.21 Sequence activities
	6.4 Estimate activity durations	4.3.22 Estimate activity durations
	6.5 Develop schedule	4.3.23 Develop schedule
4	7.2b Estimate costs	4.3.25 Estimate costs
	7.3c Determine budget (for final budget decision)	4.3.26 Develop budget
5	8.1c Plan quality management	4.3.33 Plan quality
6	9.1c Plan resource management	–
	9.2 Estimate activity resources	4.3.16 Estimate resources
	9.3c Acquire resources	4.3.17 Define project organization
7	10.1c Plan communication management	4.3.38 Plan communications
		4.3.39 Distribute information
8	11.2 Identify risks	4.3.28 Identify risks
	11.3 Perform qualitative risk analysis	4.3.29 Assess risks
	11.4 Perform quantitative risk analysis	
	11.5 Plan risk responses	4.3.30 Treat risks
9	12.1c Plan procurement management	4.3.35 Plan procurement
	12.2a Conduct procurements	4.3.36 Select suppliers
10	13.3a Manage stakeholder engagement	4.3.10 Manage stakeholders

The related standard processes are:

PMBOK	ISO 21500
4.2c Develop project management plan	4.3.3 Develop project plans

See Appendix A1 for details.

5.1.1 Critical path model integration

All activities for **phase 5 execution** must be integrated into the critical path model so that total duration, cost, and resources can be estimated. Thus the work breakdown structure (WBS), organization breakdown structure (OBS), and other breakdown structures need to be all inclusive.

5.1.2 Inclusion of capital projects in a turnaround

The inclusion of major maintenance and engineering work in a turnaround is processed within the normal turnaround framework (see Box 4.9 for an example). However, capital projects that tie into the existing plant need parallel development and execution processes (see Box 5.2).

The normal process is for tie-ins to the capital project to be made during a turnaround or online with appropriate isolation. The capital project is then built on an adjacent site and either commissioned with the existing plant online or at the end of the following turnaround or, alternatively, at the end of a shutdown that has occurred for other reasons.

Example

In the 1980s, Chevron Cape Town Refinery installed a waste heat boiler (WHB) in the flue of a fluidic catalytic cracking unit (FCCU). Tie-ins to the main flue, steam, and feed-water systems were made during a turnaround. The WHB was built while the FCCU was online between turnarounds and commissioned at the start-up of the following turnaround.

The tie-ins were work packages processed within the normal turnaround scope development.

Alternatively, process tie-ins could be done online using hot tapping, but the risk of doing this while the plant is operating is higher than it would be during a turnaround. Box 5.1 considers the debate for including tie-ins in a turnaround or carrying out hot taps while the process is online.

The commissioning and start-up of a new plant are separate, but parallel, activities to a normal start-up process. This is discussed in Section 7.8.4 and Appendix D. The parallel process is depicted in Box 5.2.

5.2 Scope
5.2.0 Introduction and relevant PM processes

In the detailed planning phase the scope is broken down into groups of related activities. The initial validation process also takes place to ensure only those items that have to be done in the turnaround are included.

The relevant standard processes are:

PMBOK	ISO 21500
5.4 Create WBS	4.3.12 Create WBS
5.5a Validate scope	4.3.13 Define activities

See Appendix A2 for details.

5.2.1 Creating the work breakdown structure

For turnarounds, the general groupings for creating the work breakdown structure (WBS) are:

Complex.

    Process unit.

        Equipment type/group.

            Equipment number.

Box 5.1: To hot tap or not to hot tap

The basic principle for deciding what work should be included in a turnaround is "If it can be done online, leave it out of the turnaround."

A dilemma ensued when a particular job entailed a tie-in to a fuel header that could not be isolated while the refinery was online. Two approaches were proposed.

1. Hot tap the header with the refinery online
2. Tie in the line to the fuel header during the turnaround.

Naturally, the hot tap entailed **higher risk**, but would keep the plant going. Alternatively, completing the tie-in as part of the turnaround could extend the schedule and time that the plant was out of action.

The issue was resolved after assurances were obtained that adequate resources were available and, critically, that there would be no effect on the critical path of the work. The work was subsequently completed during the turnaround.

Lesson learned

In this case the **risk factor** for doing the work online was part of the decision process.

This process is not always black and white and sound judgment is required when making decisions of what work should be included in a turnaround.

Box 5.2: Parallel capital project and turnaround development

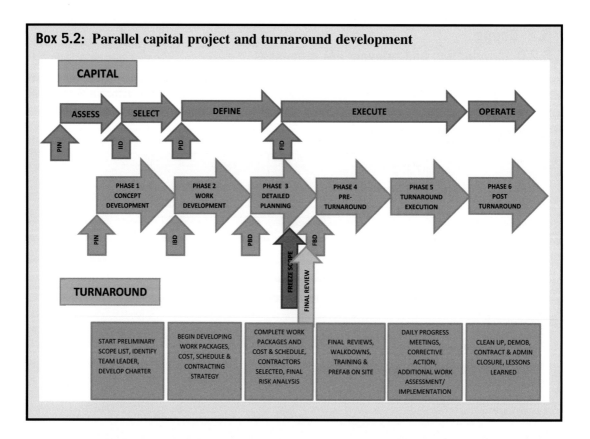

At process unit level, which is generally a subproject, code levels are entered into the critical path method (CPM) software to create the WBS.

Example

Level 1:	3 Digits	Plant Number	004	4 Crude Distillation Unit (CDU)
Level 2:	4 Digits	Equipment Type	004E	4 CDU Exchangers
Level 3:	5 Digits	Equipment Group	004EA	4 CDU Atmospheric Overhead Exchangers
Level 4:	10 Digits	Equipment Number	004EA0221A	4 CDU Atmospheric Overhead Feed Exchanger

Typical equipment groups for a refinery or gas plant are:

- Columns and vessels
- Reactors
- Exchangers
- Furnaces
- Pumps
- Compressors and turbines
- Lines and fittings
- Electrical
- Instruments

Further details and examples are given in Appendix C2.

5.2.2 Finalizing the work list (validating the scope)

Finalizing the work list so that it can be relied upon to make accurate estimates is one of the most important tasks the core team undertakes during this phase.

The **preliminary work list** must incorporate all requested work and comments for inclusion or exclusion in the turnaround.

Some work listed on the preliminary work list would require engineering input and thus an **engineering work order (EWO)** would need to be initiated. An EWO log should be maintained to chase engineering work required for the turnaround. This EWO may need to go through a **management of change** process. The resulting turnaround work is "turned over" in a **construction turnover package**, including drawings and specifications, to the core team for inclusion in the **final work list**. Box 5.3 shows a format for the EWO log.

The **final work list** must contain all work to be performed in the turnaround, including reference to any EWOs. Originators of items that were on the Preliminary Work List but that are now excluded from the **final work list** need to be notified of the decision and reasons for the decision. An example of a **work list** is shown in Box 3.6. The **final work list** will be the most up-to-date revision.

A meeting should be set up with all interested parties to agree on the **final work list**. It is then signed off/approved by the core team and the decision executive before being distributed. **Once the final**

Box 5.3: Engineering work order log format example

2002 FCCU turnaround EWO log													
EWO no	Req no	Shop Order	Item	Area	Equip no	Job & description	Engr	Contact	Due Date	Date Rec	Drafting	MOC	Comment
B301E1	BK301	2840-XE	100	Frac	D-100	Repair C100 trays 3 & 4 and replace tray 2	rbh	pwh	1/3/2002	1/6/2002	Y	N	BoM for quotes issued
B927E1	BH901	3050-20	300		Instr	Replace Cu instr tubing with galv steel	djh	ehc	1/5/2002	1/7/2002	Y	Y	MOC meeting scheduled

---

work list has been approved, it may no longer be amended. Any additional work has to go through the change control procedure described in Section 7.1.4.

Prior to finalizing the work list, a rigorous final scope challenge exercise should be carried out. This could entail a work scope peer review by an independent peer review team (IPRT) or a scope challenge workshop with turnaround project members and independent members. Shell Global Solutions offers this sort of scope challenge workshop, outlined in Box 5.4.

Box 5.5 shows the application of a Shell scope challenge workshop in a gas plant.

This clearly demonstrates that the outcome of the scope challenge exercise could result in a large reduction in agreed turnaround work.

Example Scope Challenge Outcome

- *Total work items: 498*
- *Total items approved: 250*
- *Denied items: 248*
- *Deferred items: 44*
- *Shifted items: 93*

5.3 Schedule
5.3.0 Introduction and relevant PM processes

All detailed planning is completed in phase 3 and results in a complete picture of all turnaround activities. They are sequenced to determine overall resourcing by the hour for each type of resource, as well as the final estimates of cost and duration for the entire turnaround.

The standard processes are:

PMBOK	ISO 21500
6.3 Sequence activities	4.3.21 Sequence activities
6.4 Estimate activity durations	4.3.22 Estimate activity durations
6.5 Develop schedule	4.3.23 Develop schedule

See Appendix A3 for details.

Box 5.4: Shell scope challenge workshop

Workshop objectives

1. Provide an external perspective on the turnaround scope
2. Balance cost reduction and operational integrity
3. Ensure application of risk assessment techniques in turnaround scope development
4. Obtain final agreement on turnaround scope—core team ownership

Methodologies

- Risk assessment matrix challenge process
  - Use risk assessment techniques to support decision-making of scope item
  - Apply the RAM decision matrix
- J-factor process
  - Apply the J-factor process to quantify cost and risk impact of possible decision solutions (mitigation based on cost to benefit ratio)

Challenge process

5.3.1 Turnaround model

The planners convert the final work list into activities in the turnaround model from which the final critical path analysis and resourcing is developed.

The model integrates sequencing and resourcing for each activity as well as estimated activity durations.

> **Box 5.5: Application of Shell scope challenge workshop**
>
> A gas plant typically had turnarounds in excess of 5 weeks, which was way above the industry norm. This was unacceptable to management.
>
> Shell Global Solutions (SGS) was called in to facilitate a series of "scope challenge" workshops, and, after vigorous excising of items on the work list, the turnaround plan was reduced to 3 weeks. As it happened, the actual duration turned out to be even less than the agreed plan.
>
> Consequently, turnaround staff who had been trained in these workshops went on to advise other gas plants within the company on the optimization of turnarounds, which resulted in improvements of turnaround durations elsewhere in the company.

Critical path method

A critical path network shows logical relationships between activities. The CPM establishes this network using a single estimated time (duration) for each activity.

The turnaround model was introduced in Section 1.5.1 and the development of the model starts by establishing the following:

1. Work breakdown structure
2. Organization breakdown structure
3. Calendars of working hours and days
4. Resource codes and availability

A critical path network is then built, using a copy of the network from the previous turnaround of the same process unit, if possible. Once the network has been built, critical path analysis is carried out. This is repeated many times as the model continues to be developed over the months before the turnaround. Once the work list has been approved, the critical path is finalized, including completion of resource leveling, until it is finally frozen. This is referred to as "setting the baseline." The baseline is the approved plan from which deviation is measured.

The planners use the data from the model in conjunction with the materials purchase commitments from the computerized maintenance management system to produce a *cost estimate* for inclusion in the final expenditure request.

Resource leveling

Resource leveling is required to allow the scheduling tool to calculate the number of tradespeople required per day. The number of people on-site should be minimized, but an adequate number of tradespeople are required to ensure that the critical path, subcritical paths, and the overall bulk work are achieved in the agreed time frame. Emergent work that is likely to be required also needs to be factored in.

Resource leveling is generally completed in *two stages*:

1. The *first* is done in the **concept development and work development phases** (see Sections 3.6.7 and 4.6.1). In these phases the contractors have not yet been identified and thus the resources are generic.
   For example, a welder is obtainable from any source for work anywhere.

2. The *second* is done in the **detailed planning phase** when specific contractors are identified. This helps with allocation of work areas to various contractors based on their available resources and expertise.

   For example, a welder is required from contractor X for work in area Y.

There are two resource leveling modes:

1. Resource limited mode when the software will level resources within the slack (float) time and attempts to maintain activity resourcing within the limits of availability and possibly push out the completion date of the project.
2. Time-limited mode when the software will level resources within the slack (float) time but will exceed the resource limits when unable to use the slack time to avoid going beyond the set critical path end date.

Whenever resource leveling is performed, the ensuing steps are generally followed:

1. Set start date for the turnaround
2. Allocate resources to all activities
3. Set up/update resource allocation tables with type of resource, number available, and periods available for each resource
4. Run the critical path analysis and note total critical path duration and completion date
5. Do resource leveling in resource limited mode (not time constrained) and note change in completion date (if any)
6. Do resource leveling in time-limited mode (time constrained) and note the resource overloads
7. Attempt to reduce peaks by rescheduling, splitting, stretching, and/or reprofiling activities
8. Add a schedule contingency allowance (discussed in the next paragraph)
9. Repeat steps 2–8 again until satisfied that the model is in line with the turnaround philosophy.

   Example

   A crane is allocated to lift a heat exchanger into place. Critical path analysis determines that this activity could be done at midnight. However, in the resource allocation table, this crane has been entered for use only during daylight hours. (The resource has its own calendar separate from the activity calendar, which allows work around the clock.) Consequently, the software will automatically schedule the work for sunrise the following day.

Activity prioritization
Activity prioritization determines which activities have preference when resources are restrained. If activities are scheduled in parallel, high-priority activities are scheduled first. Activity prioritization assists with obtaining access to the internals of equipment to determine potential emergent work as early as possible in the turnaround. This is discussed in detail in Appendix C5.

5.3.2 Contingencies
A *contingency or contingency reserve* is a separately planned quantity used to allow for future situations that might have been only partly planned for.

Example

The planning allocation allows for the replacement of three trays in a particular distillation column. However, when the column is opened, it is found that eight trays have to be replaced. This increases both time and cost, and this could then be covered by the contingencies for schedule and cost.

If, however, opening the column was not initially planned, but during the turnaround a decision is made to open it, this would be regarded as a change of scope and must be channeled through the change control process.

Estimating contingency is an art form. It requires experience of the process and knowledge of occurrences in previous turnarounds. Cost and schedule contingencies are interrelated, but in some cases it might be appropriate to do separate contingency calculations, one for cost and another for schedule, and then review the impact of one on the other.

Schedule contingency

There are a number of ways to determine schedule contingency. Some are:

- Area contingency
- Equipment type contingency
- Network path contingency
- Project contingency at the end of the critical path
- Contingency determined by resource leveling
- Contingency determined by critical path risk assessment
- "Fat" built into each activity (not recommended as excessive contingency can accumulate).

Determining contingency from critical path risk assessment can be complex. High-risk items on or near the critical path are going to require a much greater contingency than low-risk items. As allowed in some software packages, the use of PERT durations (low, most likely, and high) for activities of high risk, and a single duration for activities of low risk, can help with determining contingency.

Example

*In the case of the distillation column, where there is a great deal of uncertainty about the findings when opening the column, historical records can be reviewed and the **shortest, most likely, and longest durations** can be estimated. Some software will allow entering of these durations in the model to determine a **mean expected duration** for column repair with single duration values for each of the other activities, such as opening and boxing up the column (see Appendix C9).*

5.3.3 Schedule optimization

This is the most important turnaround process prior to freezing the approved work for the turnaround, as built into the critical path model.

The core team and the main contractor's management team need to be totally involved and fully conversant with the behavior of the model. All activities for the particular turnaround must be included in the model.

Schedule optimization includes:

- A thorough evaluation of the critical path and near critical path activities
- An assessment of the relationship between the number of near-critical path activities and the probability of achieving the planned completion date
- Probability analysis of **high-risk activities** and the influence on the completion date.

Schedule optimization process: outcome example

- *Reduce turnaround duration from 34 days to 27 days*
- *Identify*
  - *Critical path*
  - *Near-critical paths*
  - *High-risk activities*
- *Smooth resource peaks.*

Schedule optimization is discussed in detail in Appendix C6. Shell Global Solutions offers a schedule optimization workshop.

5.3.4 Detailed planning of major maintenance projects

Close attention to integration with routine turnaround work is required when planning major maintenance projects. Generally, this should not affect the overall duration of the turnaround where this work is likely to be done in parallel. Box 5.6 describes such activities.

5.3.5 Freeze date and baselining

The freeze date is set when all accepted work, as per the final work list, is integrated into the critical path model. No further work is accepted unless it is processed through the change control process described in Section 7.1.4. This is done at least 6 weeks before start of the turnaround execution phase so that contractual arrangements can be finalized.

As described in Section 5.3.1, the critical path model is built from the final work list. It is then massaged as described in Section 5.3.3 to achieve a duration, resource loading, and cost acceptable to all interested parties. The model is then saved as the baseline for measuring progress. The baseline is the final planned schedule for the turnaround, and any approved additional work that is added to this model may not change this baseline. Thus actual progress and cost, as well as additional work and related costs, are all measured against this baseline.

5.4 Cost
5.4.0 Introduction and relevant PM processes

The final cost of the turnaround is determined in phase 3. This cost is based on a detailed critical path model that is frozen once the final work list has been agreed and approved. The final budget is submitted to the decision executive for approval, together with the related scope, schedule, risk profile, etc.

Box 5.6: Parallel major maintenance project planning

The replacement of FCC cyclones was planned for the next turnaround (see Box 4.9 *Method 2*) and the planning staff agreed that replacement of the top of the reactor, with the cyclones attached, was the best approach.

The order was placed for manufacture of the cyclones and reactor top by a reputable manufacturer in Canada. The contract for fitting the replacement cyclones was awarded to a competent contractor who had a lot of experience and was carrying out one replacement per year on average at that time. The replacement activities were planned on Primavera and the duration had been honed down to 30 days, which included a schedule contingency allowance.

The contractor indicated that they would be able to work parallel to other FCC turnaround activities and thus the overall duration of the turnaround would not be affected. At that time, the average duration for entire FCC turnarounds was 42 days (see Box 2.9). The replacement work was completed in the agreed time frame, but the actual total turnaround duration was well beyond industry norms. However, the incompetent company planners were provided with a useful scape goat when questioned.

Comments

In hindsight, it might have been better to give responsibility for planning and supervising the entire turnaround to the FCC cyclone contractor who had extensive experience.

In this case there were other turnaround management issues, similar to those discussed in Box 3.11, which prevented completion close to an average industry-related time period.

The related standard processes are:

PMBOK	ISO 21500
7.2b Estimate costs	4.3.25 Estimate costs
7.3b Determine budget (for final budget decision)	4.3.26 Develop budget

See Appendix A4 for details.

5.4.1 Cost estimating

The total cost of the turnaround has to be estimated in two primary stages.

1. Budget proposal item to obtain the initial budget decision, which must be included in the following year's business plan and budget (see Section 3.4.3).
2. Final cost, after the scope has been frozen, for the Final Budget Decision (FBD).

A preliminary budget decision may be required (see Section 4.4.1).

Final cost must be based on:

1. Total man-hours based on the frozen critical path model once the baseline has been set
2. Man-hour costs by trade at the actual time of the turnaround
3. Stock material costs from the enterprise resource management (ERM) materials module
4. Purchased materials as per purchase contracts
5. All tools and equipment costs based on the frozen CPM model once the baseline has been set
6. Contingency allowances for emergent and extra work
7. All overhead costs.

5.4.2 The cost model

Every activity must have a cost code, whereby all resources (human, materials, and tools) have costs attached to them. Items such as overheads need to be given activities that span the full turnaround. All costs in the turnaround budget must, therefore, be included in the model. In the end, each activity would be linked to a cost center and the total of all costs would be accommodated in the critical path model. Only then can actual expenditure against the baselined planned expenditure be monitored.

5.4.3 Cost contingency

There are numerous ways of calculating cost contingency, starting with a simple percentage of total cost, progressing to complex models using risk analysis.

Table 5.2 demonstrates a simple method using a risk assessment technique for determining contingency.

5.5 Quality
5.5.0 Introduction and relevant PM processes

In phase 3, detailed quality control plans are developed for each item of equipment (e.g., pressure vessels, etc.). Weld procedures are also prepared for each type of weld.

The relevant project standard processes are:

PMBOK	ISO 21500
8.1c Plan quality management	4.3.33 Plan quality

See Appendix A5 for details.

5.5.1 Quality control

Quality control is normally driven by the inspection department. Some process plants have their inspection departments certified as independent inspection authorities with respect to the inspection of pressure vessels.

Table 5.2 Risk assessment.

Description of risk event	Probability of occurrence	Estimated cost of consequence (amount at stake)	Risk event status (criterion value)
Risk event no. 1	Probability P	Cost C	PXC
Risk event no. 2	P	C	PXC
etc.			
Project estimating contingency based on:			Sum of PXC

Example

In South Africa, refinery inspection departments can be accredited by SANAS[1] under ISO 17020[2] as independent inspection authorities.

Any request for modifications to pressure systems must, therefore, be accompanied by detailed step-by-step instruction from the inspection department. This is sometimes known as a quality control plan. A sample is shown in Box 5.7. This could be the result of an inspection advice ticket issued by the inspection department.

The quality requirements of certain standards, such as ASME and API, must be strictly adhered to.

Extensive nondestructive testing is carried out before and during the turnaround. The most common tests are as follows:

- X-ray (radiography)
- Ultrasonic
- Magnetic particle
- Liquid penetrant
- Eddy current
- Wall thickness gauging
- Surface finish measurement
- Coating thickness measurement
- Pressure (hydro-test, gas test, or service test).

For oil refinery work, all welders entering the refinery site have to be on record in the inspection department, each with their own unique identification number, as a currently certified welder for the particular welding that he is going to undertake during the turnaround. Post-weld heat treatment graphs are signed off by the client's representative and filed in the equipment files. Weld tests, such as X-rays, are approved by an inspector before "boxing up" a pressurized system.

Examples

1. *A welder certified to do carbon steel welding cannot be permitted to do alloy welding, whereas a welder certified to do alloy welding is permitted to do carbon steel welding.*
2. *Some Middle East refineries require internal trade testing of all artisans working on their process plant turnarounds. These artisans are then also required to wear their qualification ID at all times while on-site (see Box 3.9).*

Hydro-tests, relief valve tests, etc. need to be witnessed by the client's representative and signed off on a manual form that is filed in the equipment records once the equipment maintenance database has been updated.

Quality incidents costing either time or money or causing danger to life should be documented in the incident management system as described in Section 5.8.4.

Box 5.7: Sample quality control plan

REFINERY INSPECTION QUALITY CONTROL PLAN

Priority 1 = immediate. Priority 2 = within 3 months. Priority 3 = within 12 months. Priority 4 = T & I

QCP Title: T&I REPAIRS		
QCP No: I-1345-96 REV 0	Equip. No: 5C-1	Priority: 4
WORE No: ___	Equip. Name: VISBREAKER FRACTINATOR	
Originator: M.J.MOLYNEAUX		Date: 24/05/96

Act No	DESCRIPTION OF ACTIVITIES	Verified by Fabricator/ Contractor	Hold, Monitor, Review
1.	NOTE: REPAIR OR MODIFICATION OF THE TOP HEAD AND UPPER 5 METRES OF THE SHELL TO BE DETAILED IN A SEPARATE QCP.		
2.	FABRICATE NEW NOZZLE LINERS AND TRIM PLATES FROM 4MM THICK TYPE 405 STAINLESS STEEL PLATE FOR INSTALLATION INSIDE FOUR OFF NOZZLES WITH TEMPORARY LINERS AT THE BOTTOM OF THE COLUMN. THE NOZZLES IN QUESTION ARE RESPECTIVELY LOCATED AT THE NORTH (ONE X 2 " & ONE X 4") AND NORTH-WEST ORIENTATIONS (ALSO ONE X 2" & ONE X 4"). THESE ARE THE RAMSHORN NOZZLES AT THE TWO HIGHEST LEVELS. SEE ATTACHED DRAWING NO. SAB 4159 REV 1 FOR NOZZLE FABRICATION DETAILS.		M
3.	REMOVE EXISTING LINERS BY GRINDING AWAY THE ATTACHMENT WELDS AND CLEAN AREAS BY GRINDING AND WIRE BRUSHING IN PREPARATION FOR INSERTING AND WELDING NEW LINERS. INSPECTION TO CHECK ON COMPLETION OF PREPARATION.		H
4.	DRILL AND TAP 8MM DIAMETER HOLES IN THE TOP WALLS OF THE NOZZLES AS DETAILED IN ABOVE REFERENCED DRAWING.		M
5.	CUT AND FORM THE TRIM PLATES TO SUIT THE CURVATURE OF THE VESSEL SHELL.		M
6.	FIT UP AND TACK WELD NEW NOZZLE LINERS TO TRIM PLATES. WITHDRAW NOZZLE LINERS SLIGHTLY THEN FULLY FILLET WELD THE LINERS TO THE TRIM PLATES BEFORE RE-INSERTING INTO NOZZLES FOR FIT UP TO SHELL. ALL WELDING TO BE IN ACCORDANCE WITH WPS NO 1-7/M/WB/34. THIS WELDING SEQUENCE IS NECESSARY TO AVOID SEALING A DEAD SPACE BETWEEN THE TRIM PLATES AND THE SHELL. INSPECTION TO CHECK WELDS BEFORE RE-INSERTING INTO NOZZLES		H
7.	FULLY FILLET WELD THE TRIM PLATES TO THE VESSEL SHELL BEFORE MAKING EXTERNAL FILLET WELDS ONTO THE NOZZLE FLANGES.		M
8.	COMPLETE EXTERNAL FILLET WELD FOR SEALING THE LINERS TO THE NOZZLE FLANGES.		M

QCP number: I-1345-96 REV 0

Document Title: Quality Control Plan Standard Form

Document Number: INSP\PRES\FOR002 - 2

Approved:

5.6 Resources

5.6.0 Introduction and relevant PM processes

Final resource planning is carried out in phase 3 where detailed estimates are derived from the critical path model.

Relevant standard processes are:

PMBOK	ISO 21500
9.1c Plan resource management	—
9.2a Estimate activity resources	4.3.16 Estimate resources
9.3c Acquire resources (contractors, tools, and equipment)	—
	4.3.17 Define project organization

See Appendix A6 for details.

5.6.1 Development of the organization breakdown structure

Resources for management are built into the OBS. Resources for carrying out the work must be included in the model and attached to each activity. This is explained further in Appendix C2.

5.6.2 Detailed planning team

Detailed planning should be performed by those with an intimate knowledge of what is to be done in the execution phase of the turnaround. It is strongly recommended that area and discipline specialists, under the guidance of the planning engineer, undertake the planning of their respective areas of expertise. Ideally, these specialists would also be the supervisors of their respective areas during the execution phase.

Depending on the contractual arrangement, these team members may be staff members or could be contractors.

To build the activity network diagrams, the planning team needs access to at least:

I. Final work list and prioritization
II. Tools and equipment lists
III. Labor categories and appointed contractors lists
IV. Materials (in materials module of ERM and direct orders lists)
V. Quality control plans and relevant drawings
VI. Cost categories in the budget.

5.6.3 Resource leveling 2

Resource leveling is built into the critical path model as described in Section 5.3.1. The outcome is a resource histogram for each trade and item of equipment (e.g., crane) showing the required level of resourcing, by the hour, throughout the turnaround. Resource histograms are explained further in Box 6.5, Box 7.4, and Box C17.

Particular attention needs to be paid to scarce resources such as alloy welders.

Example

Extensive relining of a distillation column was required using alloy welders in a particular turnaround. Since alloy welders were a scarce resource and required to work in a confined space, they were found to be on the critical path.

5.6.4 Equipment and tools

All required equipment and tools must be attached to the relevant activities in the critical path model. Over-resourcing of items that are in short supply needs to be critically reviewed when doing resource leveling.

Examples

- *Post-weld heat treatment (PWHT) rigs for heat treatment of alloy welds*
- *Ultra-high-pressure cleaning machines for cleaning those internals and heat exchanger bundles that are difficult to clean.*

5.6.5 Productivity

The critical path model will plan and record actual man-hours at the work site (on the job). The time clocking system will record arrival and departure from site. These need to be reconciled to determine the predicted total man-hours for the turnaround.

Keep in mind that once an individual clocks in, he/she needs to change clothing and obtain requisite work permits, tools, and materials before proceeding to the work site. Tea and meal breaks also need to be factored in. At the end of each shift, packing up and changing of clothing before clocking out also takes time. As a result, actual clocked man-hours and real on-the-job time could be markedly different.

One method to reconcile this is to factor the clocked time against the time on the job to determine total man-hours required (factor = time on the job/clocked time). This would require good historical data for total clocking and on-the-job time, which determines a factor for a particular site. Note that on-the-job time should be the same for identical equipment at different sites, and thus critical path models could be alike for similar plants at other sites.

Whatever approach is taken, consistency across the site and from one turnaround to the next is imperative.

The measurement of reported times is discussed in Section 7.6.2.

Some common factors that need to be taken into account when determining productivity include:

- Weather conditions
- Experience levels of the available trades people
- Quality of contractor supervision
- Concentration of work activities
- Planned shift hours and overtime (see Appendix C2 discussion on calendars)
- Plant logistics and expected delays
- Integration with capital work.

5.7 Communication

5.7.0 Introduction and relevant PM processes

Detailed planning procedures are applied in phase 3. In addition, the plot plan is drafted and approved. Scope-related meetings come to a close at the end of this phase once the scope has been agreed and finalized.

Related standard processes are:

PMBOK	ISO 21500
10.1c Plan communications management	4.3.38 Plan communications
—	4.3.39 Distribute information

See Appendix A7 for details.

5.7.1 Key procedures

All recommended procedures, checklists, etc. are listed in Appendix E. However, the key procedure when doing detailed planning relates to building and maintaining the critical path model. As discussed in Section 4.3.1, it is necessary to have standard codes, calendars, etc. This procedure would lay the ground rules for building the critical path model. An example of the content is shown in Box 5.8.

5.7.2 Flawless turnaround awareness program

This is also planned in phase 3. The goal of this program is to promote the minimization of incidents related to health, safety, environment (HSE), and quality. The number of incidents is also a key performance indicator for every turnaround.

5.7.3 The plot plan

A copy of the original plant layout plot plan of the complex is a suitable start, and turnaround items are superimposed on this copy. In some cases a Google Earth view may be acceptable on which to overlay turnaround items.

Box 5.8: Sample turnaround planning procedure contents

Table of contents

The original plan would include a view of all plant, equipment and piping runs, roads, and buildings. Items to be superimposed include:

1. Lay-down areas for new equipment that is going to be installed
2. Tool sheds (usually ISO containers) with all tools required, in areas that are in close proximity to the relevant plant
3. Small materials and parts stores (again, usually ISO containers) for all bolts, gaskets, etc. again in areas that are in close proximity to the relevant plant. These would have been reserved in the main store during the planning phase, and then relocated to this store.
4. Road routes and access routes clearly showing crane access, general site vehicle access, or no access at all. All private vehicles and buses are to be kept off-site. Crane access needs to be very specific showing nonload-bearing areas. Road plates may be required to protect certain access routes. A clear indication of the direction of traffic may be required in certain congested areas.
5. Lay-down areas for hazardous substances such as catalysts, chemicals, lubricants, and welding gases
6. Lay-down areas for skips for contaminated items and refuse

7. Lay-down areas for redundant/scrap equipment/plant
8. Marshaling areas for scaffolding and insulation contractors
9. Parking areas for welding machines, air compressors, generators, and fuel tenders
10. Parking areas for cranes and heavy equipment
11. Parking areas for site vehicles
12. Areas for offices, mess rooms, change rooms, and toilets (with the project control office clearly indicated)
13. Temporary cabling and piping routes
14. Emergency assembly points
15. Eye wash facilities
16. Smoking areas
17. First aid rooms.

The plot plan is to be approved by the turnaround project manager and other relevant parties before being printed and displayed in prominent areas of the complex. Box 5.9 gives an example of a simple plot plan.

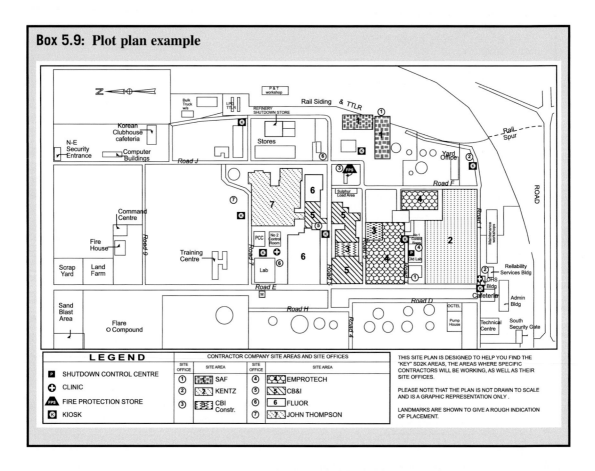

Box 5.9: Plot plan example

5.7.4 Meetings

Scope review meetings are intensified, resulting in a final agreed scope of work. Resourcing, safety/risk, and procurement issues arising from the firming up of the scope are discussed and resolved in relevant meetings.

5.7.5 Sponsor communication

Communication with the sponsor entails presentation of the final turnaround plan and budget with related issues in order to obtain the final budget decision.

5.8 Risk (including HSE issues)
5.8.0 Introduction and relevant PM processes

In phase 3 a risk register is established, and each potential risk for the forthcoming turnaround is registered and assessed. Risk treatment is proposed, action is recommended, and mitigation for each risk is taken at the appropriate time.

The audit at the end of this phase determines the readiness to proceed with execution of the turnaround.

Related standard processes are:

PMBOK	ISO 21500	ISO 31000 risk management
11.2 Identify risks	4.3.28 Identify risks	1. Establish the context
—	—	2. Communicate and consult
11.3 Perform qualitative risk analysis 11.4 Perform quantitative risk analysis	4.3.29 Assess risks	3. Risk assessment 　Risk identification 　Risk analysis 　Risk evaluation
11.5 Plan risk response 11.6 Implement risk responses	4.3.30 Treat risks	4. Risk treatment

See Appendix A8 for details.

5.8.1 Work list contingency review

What is likely to go wrong? It is highly recommended that a contingency review be performed for all equipment to be worked on. Inspection records should be reviewed again and again, and materials contingencies developed, especially those with long-lead times. Risk is assessed using the risk assessment matrix (RAM) in Box 4.7 or other suitable means. Box 5.10 gives an example of a work list contingency review showing the associated levels of risk.

5.8.2 Risk register

A turnaround risk register is mandatory with risk items from Section 4.8.3 required to be recorded in this register. The risk register from the previous turnaround could be used as a template.

Box 5.10: Work list contingency review example[3]

	Equipment Number	Work Order Number	Job	Issue	Like	Con	Risk (Tier	Action Plan	Plan Owner
	A	B	C	D	E	F	G	H	I
3	A34C1	6007313	Replace tray #20 and #33 support beams.	materials availability			0	get drawing and check no weld to vessel is required	Materials co
4	A34C1	6007345	Water Test/Adjust Liquid Distributors	Distributer blockage in nozzle	3	3	Tier 4	contingency to clean nozzles	planner
5	A34C1	6007345	Water Test/Adjust Liquid Distributors	No people in column during water test	3	3	Tier 4	WMS and plan to reflect	planner
6	A34C1	6007344	Replace Column Packing In-Kind	as below					
7	A34C1	6007636	Insert weir plates	TSD to give details / dwg			0	TSD to give details / dwg	L.Sinclair
8	A34C1		Renew vertical ladder/cage on 34C-1	OTR - schedule before shut or do during shut			0	OTR - schedule before shut or do during shut	R.Harden
9	A34C1	6005756	T&I TASKLIST - A34C1	Packing availability	2	4	Tier 3	order new packings	Materials co
10	A34C1	6005756	T&I TASKLIST - A34C1	tray hardware	2	4	Tier 3	tray contingency or consignment container on site to suit	Materials co
11	A34C1	6005756	T&I TASKLIST - A34C1	pyrophorics or chemical residue	4	4	Tier 6	WMS & plan to include PPE for protection against chemical residue	planner
12	A34C1	6005756	T&I TASKLIST - A34C1	Discovery of damaged nozzles	3	4	Tier 5	UT gauging scope and access required pre shut	Paul Zeidan
13	A34C1	6005756	T&I TASKLIST - A34C1	coke cleaning at vapour line	2	4	Tier 3	add contingency	planner
14	A34C1	6005756	T&I TASKLIST - A34C1	Generally uns column needs a detained cleaning plan			0	created cleaning plan	planner
15	A34C1	6005756	T&I TASKLIST - A34C1	check capstone record for details			0		Paul Zeidan
16	A34C100A	6007890	Repairs to inner gas tubes on 34C-100A/B	likely damage to gas tube and refractory	2	3	Tier 1	Constructability on gas tube repairs and refractory repairs	richard
17	A34C100A	6007890	Repairs to inner gas tubes on 34C-100A/B	Material for gas tube not available	2	3	Tier 1	Utilise material in Clean Fules Yard	Materials co
18	A34C100A	6007890	Repairs to inner gas tubes on 34C-100A/B						
19	A34C100A	6007890	Repairs to inner gas tubes on 34C-100A/B	cat in 100b and maybe in 100a				plan for cat removal	planner
20	A34C100A	6006635	T&I TASKLIST - A34C100A	No issues			0		
21	A34C100B	6007071	Renew level gauge spool of C-100B	No issues			0		
22	A34C100B	6006636	T&I TASKLIST - A34C100B	No issues			0		
23	A34C2A	6006715	Renew internal tray sections	Material not available	2	4	Tier 3	Order material	scott
24	A34C2A	6007175	Gritblast & paint externally to system I	Work front conflictd due to noise / cleanup / dust / etc	2	4	Tier 3	WMS and plan to reflect	Kyle
25	A34C2A	6006083	T&I TASKLIST - A34C2A	A					
26	A34C2B	6006983	Renew 2 off vessel nozzles, B8 & B6	testing requirment not detailed in SWO	2	4	Tier 3	detail testing requirements	Paul Zeidan
27	A34C2B	6007496	Renew tray valves	Valves not available during T&I causing delaysd	2	4	Tier 3	consignment or order 100%	Materials co
28	A34C400	6007158	Gritblas/Paint vessel 34C-400/401	Area to be grit blasted not properly defined			0	Defined priority area to be grit blasted and painted	Paul Zeidan
29	A34C400	6007865	Renew N10 & N11 nipples on 34c-400/401	Nozzle N10 listed as otugged on genreral			0	Review is reapir to N10 is required	Paul Zeidan
30	A34C400	750006931	Anticipate renewal of top manway on 34C-400	testing requirment not detailed in SWO	2	4	Tier 3	Define what testing is required.	Paul Zeidan
31	A34C400	750007441	Renew top and middle manway nozzles	Bottom manway - possible PWHT and column	1	3	Tier 1	Constructability	richard
32	A34C400	50006930	renew all trays and downcomers 34C-400	awaiting approval - order materials			0	awaiting approval - order materials	scott
33	A34C400	50006919	Renew skirt fireproofing on 34C-400/401	Corrosion on skirt	2	4	Tier 3	Contingency in plan for mechanical repair	planner
34	A34C400	6007461	T&I TASKLIST - A34C400	Vessel not sufficently clean	2	4	Tier 3	Contingency in plan for hydro jetting column	planner
35	A34C400	6007461	T&I TASKLIST - A34C400	Demister pad damaged	3	4	Tier 5	Review materials	Materials co
36	A34C401	6007462	T&I TASKLIST - A34C401	Benzene detected			0		
37	A34C402	6007484	Renew internal trays	Tray hardware not available	2	4	Tier 3	Review materials requirements	Materials co
38	A34C402	6007463	T&I TASKLIST - A34C402	Vessel not sufficently clean	2	4	Tier 3	Contingency in plan for hydro jetting column	planner
39	A34C402	6007463	T&I TASKLIST - A34C402	Demister pad damaged	3	4	Tier 4	Review materials	Materials co
40	A34C406	6007926	Renew skirt fireproofing on 34C-406	Corrosion on skirt	2	4	Tier 3	Contingency in plan for mechanical repair	planner
41	A34C406	6007926	Renew skirt fireproofing on 34C-406	Corrosion on skirt	2	4	Tier 3	Include painting in plan	planner
42	A34C406	6007927	Renew B6 20 mm nipple in 34C-406	Requires PWHT			0	Include PWHT in plan	planner
43	A34C406	6007464	T&I TASKLIST - A34C406	Demister pad damaged	3	4	Tier 5	Review materials	Materials co
44	A34C411X	6007465	T&I TASKLIST - A34C411X	New vessel in 2006, no issues expected			0		
45	A34C414	6007928	Machine top manway gasket face 34C-414	Working at heights	3	4	Tier 5	Review in WMS	safety co
46	A34C414	6007467	T&I TASKLIST - A34C414	Vessel not sufficently clean	2	4	Tier 3	Contingency in plan for hydro jetting column	planner

A sample showing possible risks is shown in Table 5.3, all of which have occurred previously. The examples are shown in Box 5.11.

5.8.3 Audits and reviews

The most important audit/review takes place at the end of the detailed planning phase. Two alternative formats are described here.

Front-end loading (FEL)

A FEL index is a useful tool to assess performance in preparation for the turnaround. This index is defined as:

A quantitative measure of the level of definition for a turnaround.

Turnaround definition includes defining:

- **what** will be done,
- **how** it will be done,
- **who** will do it, and
- **when** it will be done.

Table 5.3 Sample risk register.

Initiator/ controller/ owner/actioner	Category high/ med/low	Subject area	Ref no	Risk description	Mitigation	Due date	Residual risk/closure
		Quality		Chemical cleaning: incorrect methodology[a]	Competent chem cleaning contractor appointed.		Contractor working to approved methodology.
		Schedule		Column tray collapse[b]	PERT analysis for planning.		Anticipate the worst, hope for the best.
		Schedule		Reactor catalyst removal: jackhammers may be required[c]	Arrange for jackhammers. Identify alternatives.		
		Schedule		Number of near critical activities[d]	Near critical activities assessed, CPM model optimized.		Reduced near critical activities.
		Schedule		High intensity activities: Working too close to each other[e]	CPM model optimized.		Resources leveled.
		Schedule, scope		Engineering work: Significant, complex, etc.[f]	Risk assessment of engineering work.		Reduced engineering work in turnaround.
		Schedule, quality		PWHT rigs: time and quantity[g]	Maximize pre-turnaround PWHT. Optimize use of PWHT rigs.		Use of PWHT rigs optimized.
		Resource		Alloy coded welder shortage[h]	Maximize pre-turnaround alloy welding. CPM model optimized.		Resources leveled.
		Resource		Shortage of very high-pressure cleaning equipment[i]	CPM model optimized. Investigate alternative cleaning methods.		Resources leveled
		Resource		Welder strike[j]	Union agreement or obtain nonunion welders		

Superscript[x] refers to Box 5.11 for examples.

Box 5.11: Incidents included in Table 5.3 risk register sample

This is a list of actual experiences.
a. Box 4.6: Chemical Cleaning Gone Wrong
b. Box 4.4: Collapse of Column Trays
c. Box 4.5: Liquid Carryover
d. Box 5.15: Critical Path Modeling (CPM) Buy-in
e. Section 5.6.3: Resource Leveling 2 Example
f. Box 4.9: Alternative Methods for Replacing FCCU Cyclone Replacement
g. Box 6.10: Post-Weld Heat Treatment
h. Section 5.6.3: Resource Leveling 2 Example
i. Section 4.3.5 item 3b: Very High Pressure (VHP) Heat Exchanger Cleaning
j. Box 7.7: Welder Strike.

Ratings are suggested as follows:

1. NO EVIDENCE—zero
2. POOR—25%
3. FAIR—50%
4. GOOD—75%
5. BEST—100%

Each area of assessment shown below is given a rating:

Area A: scope definition

1. Turnaround goals determined
2. Stakeholder buy-in obtained
3. Work lists for maintenance, inspection, and capital projects completed
4. Scope control measures established
5. Work orders/engineering packages completed.

Area B: execution strategy

1. Roles and responsibilities defined
2. Contracting strategy agreed
3. Schedule development completed
4. Risk mitigation and contingency planning actioned
5. Lessons learned incorporated.

Area C: planning status

1. Turnaround start date set
2. Availability of resources established
3. Materials deliveries on schedule
4. Turnaround, start-up, and operations procedures completed
5. Detailed plans, job plans, rigging plans, and scaffolding plans completed.

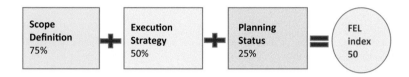

FIGURE 5.1

Front end loading (FEL) example.

Alternatively, these topics could be included in a more detailed readiness review (see below).

FEL is graphically described in Fig. 5.1.

The FEL approach has been successfully used by both Chevron and Sasol.

Readiness reviews

These are the most important reviews for a turnaround. The format could include either the FEL subject headings, or those used by Shell, and the team responsible for conducting it is sometimes referred to as an IPRT. A team of experts should conduct it, being either company representatives or outside experts independent of the ongoing turnaround planning effort. The final work scope and preparedness should be reviewed and certified by the IPRT before submission and approval by the decision executive.

A formal readiness review has been developed by Shell. This is described in Box 5.12.

Box 5.13 gives an example of the outcome of a Shell turnaround readiness review.

5.8.4 Incident management

Risk may relate to an opportunity or loss, where an incident relates to loss only.

Definition

An incident is an unexpected, undesired event that results in or has the potential to cause adverse consequences.

A **major incident** is an event that results in or has the potential to result in:

1. a fatality,
2. a disabling injury resulting in more than a few days off,
3. an environmental emission in excess of a stipulated amount, and/or
4. a financial loss in excess of a stipulated value that is regarded as unacceptable by top management.

Incidents can occur in any of the following categories or any combination of them:

1. safety
2. health
3. environment

Box 5.12: Shell turnaround readiness review

Objectives

- To arrive at an independent opinion on the preparation activities performed by the turnaround execution team and associated contractors in readiness for the turnaround
- To highlight areas successfully covered and in place for turnaround implementation
- To identify weaknesses in relation to the turnaround preparation that may lead to unacceptable risk, lost business opportunities, or wasted resources
- To provide suggested improvements to any weak areas and seek agreement on corrective actions.

Methodology

The turnaround readiness review (TRR) is designed to assess the preparedness of an organization for the pending turnaround against the background of global best practices, benchmarks, and 11 turnaround key success areas.

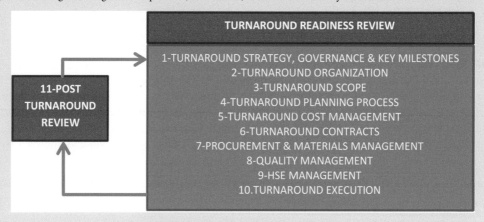

1-TURNAROUND STRATEGY, GOVERNANCE AND KEY MILESTONES: Management processes, strategy, and premises
2-TURNAROUND ORGANIZATION: Roles, responsibilities within the turnaround team
3-TURNAROUND SCOPE: Work scope development, control, and optimization
4-TURNAROUND PLANNING PROCESS: Process and preparation milestones, operations planning
5-TURNAROUND COST MANAGEMENT: Budgets, cost estimating, and control processes
6-TURNAROUND CONTRACTS: Strategy and contracting status
7-PROCUREMENT AND MATERIALS MANAGEMENT: Materials, warehouse, and processes
8-QUALITY MANAGEMENT: Flawless turnaround, planning, controls, and responsibilities
9-HSE MANAGEMENT: Planning, controls, and responsibilities
10-TURNAROUND EXECUTION: Procedures, processes and work practices, pre-turnaround preparations
11-POST TURNAROUND REVIEW: Evaluation process, lessons learned.

TRR process

- Data collection
  - Interviews with key players
  - Documents review
  - Field visits
- Analysis of data
  - Document observations on level of readiness based on data collection
  - Create scorecard that provides qualified assessment of readiness
- Conclusion and recommendations
  - List of recommendations to address findings
  - Report and gain agreement on proposed recommendations

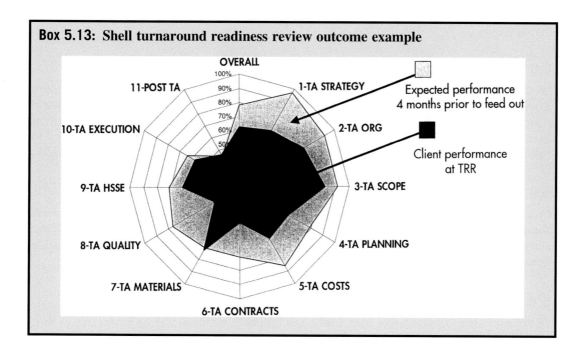

Box 5.13: Shell turnaround readiness review outcome example

4. quality
5. lost production (late start-up after a turnaround).

It goes without saying that major incidents are to be avoided, if at all possible, at all times, but particularly in a turnaround.

An incident management system needs to be established (if one has not already been established for normal operation of the plant) so that all incident events are logged and a process of corrective action is initiated to minimize the chances of a reoccurrence.

5.9 Procurement
5.9.0 Introduction and relevant PM processes

Detailed resourcing takes place in phase 3 and all required contracts for turnaround execution are placed. All materials procurements are also placed.

Relevant standard processes are:

PMBOK	ISO 21500
12.1a Plan procurement management	4.3.35 Plan procurements
12.2a Conduct procurements	4.3.36 Select suppliers

See Appendix A9 for details.

5.9.1 Contract management

The change process needs to be an integral part of contract management. The procedure for changes to a contract must be agreed upon at the negotiation stage by all parties to ensure speedy processing of changes before events overwhelm the parties involved. Major disputes have been known to continue for a long time after the turnaround has been completed if this is not done. The steps described for change control (see Section 7.1.4) must be applicable to, and included in, all turnaround contracts.

5.9.2 Contractor resource requirements

Contractors require a number of months to source the various skills required for the turnaround. Adequate notice of the numbers required for each skill, and man-hours, needs to be provided to each contractor. Estimates are normally deduced from an updated Critical Path Model from a previous turnaround.

5.9.3 Model tendering (bidding) process

To optimize the governance, risk, compliance, and assurance elements of good governance, a structured, transparent tendering system in line with international best practice is required. The basics are:

1. Decide on a suitable contracting strategy for the particular scope of work
2. Invitation to bid package to be sent out with reasonable bid period to ensure a comprehensive and complete bid submission by all participants
3. Tenders to be sealed separately (technical and commercial) and submitted into a sealed tender box by a set time on a set day
4. The sealed tender box is to be opened, and all bids registered. At this stage only technical bids should be opened and commercial bids kept sealed
5. Technical bids are to be evaluated by experts and technical approvals submitted to the multidisciplinary tender committee. This tender committee only then opens technically approved commercial bids
6. Technically approved commercial bids are to be evaluated by experts and commercial approvals and recommendations for award submitted to a multidisciplinary tender committee before obtaining line management approval (approving authority).

It is now common practice for the above process to be automated within the company ERM system with electronic submission of bids. Fig. 5.2 outlines the process.

The type of contractual arrangement and its relative risks are discussed in Section 3.9.4.

5.9.4 Award of contracts and placing of orders

As detailed planning progresses, all stock materials are reserved in stores and orders for other materials are placed.

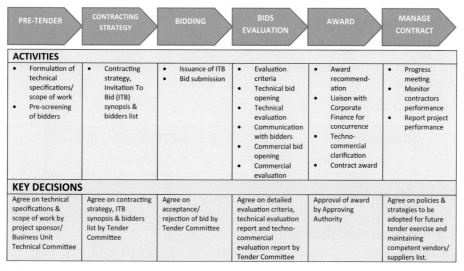

PRE-TENDER	CONTRACTING STRATEGY	BIDDING	BIDS EVALUATION	AWARD	MANAGE CONTRACT
ACTIVITIES					
• Formulation of technical specifications/ scope of work • Pre-screening of bidders	• Contracting strategy, Invitation To Bid (ITB) synopsis & bidders list	• Issuance of ITB • Bid submission	• Evaluation criteria • Technical bid opening • Technical evaluation • Communication with bidders • Commercial bid opening • Commercial evaluation	• Award recommend-ation • Liaison with Corporate Finance for concurrence • Techno-commercial clarification • Contract award	• Progress meeting • Monitor contractors performance • Report project performance
KEY DECISIONS					
Agree on technical specifications & scope of work by project sponsor/ Business Unit Technical Committee	Agree on contracting strategy, ITB synopsis & bidders list by Tender Committee	Agree on acceptance/ rejection of bid by Tender Committee	Agree on detailed evaluation criteria, technical evaluation report and techno-commercial evaluation report by Tender Committee	Approval of award by Approving Authority	Agree on policies & strategies to be adopted for future tender exercise and maintaining competent vendors/ suppliers list.

FIGURE 5.2

Tendering process framework.

All turnaround execution contracts are awarded at the end of detailed planning. Equipment hire contracts are also awarded.

5.9.5 Delivery of major equipment to site

The delivery of major equipment to site may require detailed planning and execution. Special transport and cranage, as well as government permits, may be required. Transport could include ships with heavy-lift facilities on board, barges, floating cranes, multiwheeled road transporters, and/or heavy-lift cranes (static and/or wheeled).

The transport route may need to be meticulously planned with possible strengthening of bridges and/or opening of alternative routes. Box 5.14 gives an example of the transporting and installation of a column prior to a turnaround.

Box 5.14: Delivery and installation of a distillation column

A distillation column weighing approximately 200 tons with a diameter of 1.7 m and height of 28 m, was needed to be replaced.

The column had to be lifted off a ship, using the ship's own heavy-lift cranes, onto a barge. The barge was towed by tug to a custom-made jetty where it was lifted by two land-based cranes onto a multiwheeled road transporter. (This jetty had been used previously for similar heavy equipment deliveries.) The road from the jetty to the refinery included pipe crossings and a bridge. The pipe crossings were strengthened with steel road plates and a temporary alternative route around the bridge was arranged. At the refinery the foundation for the column had been cast previously and road plates were laid down for the cranes. When the road transporter arrived at the refinery, two cranes lifted the column straight onto its foundations.

The column had been supplied with the trays in place. Walkways, ladders, and relief valves were attached and the column was insulated after erection, while the plant was on stream. All that was needed in the execution phase of the turnaround was for the connection piping to be removed from the old column and attached to the new one.

5.10 Managing stakeholders
5.10.0 Introduction and relevant PM processes

At the end of phase 3 the final proposal for the forthcoming turnaround is submitted for approval to the decision executive.

Related standard processes are:

PMBOK	ISO 21500
13.3 Manage stakeholder engagement	4.3.10 Manage stakeholders

See Appendix A10 for details.

5.10.1 Final turnaround proposal

The turnaround proposal is submitted to the decision executive as the **FBD** for approval.

It should include the following:

1. Turnaround philosophy
2. Total estimated cost for the turnaround, including contingency allowance, categorized cost breakdown, and earned value expenditure curve
3. Final work list or reference to one if it is too bulky for inclusion
4. Summary critical path analysis indicating the total turnaround duration
5. Organization breakdown structure
6. Work breakdown structure
7. Resource histograms
8. Full value of plant assets due to be shut down and cost per day of lost production for each day of the turnaround
9. List of capital work to be done in the turnaround on separate expenditure requests.

Box 5.15 gives an example of an instance where buy-in was not attained from top management with the result that necessary corrective action was not initiated, resulting in a major loss to the company.

5.11 Summary: phase 3—turnaround detailed planning
5.11.1 Focus areas

1. Finalized work list
2. Engineering turnovers issued
3. Firm budget estimate
4. Additional work procedure
5. Detailed SHEQ plan including unit preparation for entry
6. Contracting plan

> **Box 5.15: Critical path modeling buy-in**
>
> A major refinery turnaround was planned using an open-plan professional critical path model. All units were integrated into the master model including significant engineering work. Fluor, Foster–Wheeler, and CBI were the major engineering contractors, and they had prepared their plans either in Primavera or Open-Plan for consolidation into the master plan, which was managed by the operating company.
>
> After a full review, analysis, and amendment of the model, a 6-week turnaround was predicted and subsequently agreed to by management. On approval, scope was frozen and the turnaround commenced. The model was updated daily with actual progress, but after 2 weeks was predicting a start-up of 2 weeks later than planned. Management was in denial and refused to believe the results of the model. No corrective or contingency action was taken and the model continued to predict an overrun of about 2 weeks.
>
> As forecast, actual start-up was just over 2 weeks late. Since this was a regional refinery with no other refineries in the area, they had to stockpile product for the duration of the turnaround. As a result of the late start-up, the stockpile ran dry and product had to be railed in from other refineries at great cost. Unsurprisingly, the company made a major loss that year. Not only was the company affected but the fuel shortage also had a detrimental impact on the local economy.
>
> **Lessons learned**
>
> This was a complicated turnaround with extensive engineering work, raising the probability of not completing the work on time. Probability tools were not commonly used at that time.
>
> **Communication:** It was clear that management did not trust the model. The situation might have been somewhat allayed had the planners explained "the build" of the model in more detail to management before the turnaround in order to build trust.
>
> Finally, it should be noted that a situation such as this should be declared an incident, and entered into an incident register, as it resulted in a major financial loss to the company.

7. Critical path schedule and detailed work schedules
8. Purchasing plan
9. Logistics plan.

5.11.2 Deliverables

Final budget decision, including:

- Expenditure requested based on a ±10% estimate
- Final execution plan
- Audit report
- Other supporting documents.

5.11.3 Tools

1. Contracting plan
2. Expenditure request requirements
3. Critical path scheduling
4. Audit.

5.11.4 Timing

3–6 months prior to turnaround.

References

1. South African Association of Approved Inspection Authorities. Available from: http://inspectionauthority.co.za/.
2. ISO 17020 general criteria for the operation of various types of bodies performing inspection.
3. McGrath R. *The recipe for success: investing in comprehensive shutdown planning STO Asia*. IQPC; 2014.

Further reading

1. Buckner E. *Optimize turnaround projects: effective planning can eliminate surprises during shutdown and construction*. Hydrocarbon Processing; September 2005.

SECTION

C

Turnaround project execution

Phase 4: Pre-turnaround execution

6.0 Outline

The turnaround project now progresses from planning to execution and during this phase the preexecution activities of the project are carried out. These include contractor mobilization, erecting prefabrication on-site, team training, and preparation of the plant for shutdown. The final date and time for the start of execution of the turnaround is also agreed.

This is covered in the following Knowledge Areas:

1. integration
2. scope
3. schedule
4. cost
5. quality
6. resources
7. communication
8. risk (including health, safety, and environment [HSE] issues)
9. procurement
10. stakeholder management.

Specifically Table 6.1 references PMBOK and ISO 21500 processes applicable in this phase.

6.1 Integration
6.1.0 Introduction and relevant project management (PM) processes

The project moves from planning to preparation for the execution of the turnaround in phase 4.

The relevant standards are:

PMBOK	ISO 21500
4.3a Direct and manage project work	4.3.4 Direct project work

Table 6.1 Phase 4—pre-turnaround execution.

No.	PMBOK	ISO 21500
1	4.3a Direct and manage project work	4.3.4 Direct project work
2	5.6 Control scope	4.3.14 Control scope
3	6.7 Control schedule	4.3.24 Control schedule
4	7.4 Control costs	4.3.27 Control costs
5	8.2a Manage quality	4.3.33 Perform QA
	8.3 Control quality	4.3.34 Perform QC
6	9.3b Acquire resources	—
	9.4b Develop team	4.3.18 Develop project team
	9.5a Manage team	4.3.20 Manage project team
	9.6a Control resources	4.3.19 Control resources
7	10.2a Manage communication	4.3.40 Manage communications
	10.3a Monitor communications	4.3.39 Distribute information
8	11.2 Identify risks	4.3.28 Identify risks
	11.3 Perform qualitative risk analysis	4.3.29 Assess risks
	11.4 Perform quantitative risk analysis	
	11.5 Plan risk responses	4.3.30 Treat risks
	11.6 Implement risk responses	4.3.31 Control risks
	11.7 Monitor risks	
9	12.3 Control procurements	4.3.37 Administer procurement
10	13.3b Manage stakeholder engagement	4.3.10 Manage stakeholders
	13.4 Monitor stakeholder engagement	—

See Appendix A1 for details.

6.1.1 Overall management of the pre-execution phase

The pre-turnaround phase (phase 4) covers the time period immediately before the full-scale execution of the turnaround. The main focus areas are:

1. Team training and orientation
2. Mobilization
3. Final turnaround execution plan
4. Pre-turnaround work

Needless to say, the focus of the core team also moves from planning to execution.

Checklists are useful tools to ensure everything has been covered in preparation for the turnaround. These are reviewed and anything out of place or missing is rectified.

Box 6.1: Bapco T&I checklist (c.1995 adapted)

Shutdown preparation	Organization
Obtain	Prepare
1. Shutdown records	1. Trailer and equipment plan
2. EP and RC minutes	2. Temporary lighting plan
3. Mechanical work list	3. Locations for staging
4. Electrical work list	4. CTS tool list
5. Instrument work list	5. Special tools list
6. RV list	6. Requisition rations
7. Thermowell list	7. Staging list
8. Blind list	8. Organization chart
9. OPD valves list	9. Shop work schedule
10. Inspection valve list	10. BP 1600 books (work instructions/change orders)
11. Seal weld list	11. Turnover sheets
12. P&U electrical work list	12. List/schedule pre SD items
13. Material reservation sheets	13. Contract specifications 1
14. TSD exchanger cleaning requests	14. Contract specifications 2
15. Exchanger test schedule	15. Contract specifications 3
16. Seal weld repair list	16. Contract specifications 4
17. USFD staging list	17. Contract specifications 5
18. Reference drawings	18. Contract specifications 6
19. Foundation repair list	19. Hydro-test records
20. OPD shutdown and start-up plan	20. Shift turnover book
21. Reserved stock items	21. Final manpower request
22. Common stock items used in last T&I	22. Stationary
23. Pre-SD staging list	23. Unit shutting down supervision
	24. Unit shutting down manpower

Site preparation	
Install or arrange for	28. Order washed sand
1. Site SD trailers	29. Overhaul lifting equipment
2. Telephones	30. Stage equipment per list
3. Safety signs and barriers	31. Catalyst handling equipment
4. Drinking water trailer	32. Test rigs
5. Washing facilities	33. Sand bags
6. Smoking shelter	34. Fill sand bags
7. Smoking permit	35. Remove insulating metal cladding
8. Materials compound	36. Oxygen bottles
9. Sandblasting compound	37. Butane bottles
10. Air and lighting: sandblast compound	38. Bottle carriers
11. Air compressors	39. Heavy equipment bogey
12. Clean unit sewers	40. Food heaters
13. Road plates	41. OPD permit board
14. Temporary unit lighting	42. Clean pipe trench
15. Off-site lighting	43. Sand/grit blast material
16. Check 440 V	44. Steam hoses
17. Check 110 V	45. Breeze Wagons
18. Check air lines	46. Fresh air equipment for flare blind
19. Mark out toolbox area	47. Transport area
20. Mark out contractors' area	48. Wrenches
21. Power to contractors' area	49. Radios
22. Still cleaners trailer	50. Dry ice
23. Heater plug racks	51.
24. Plug blasting cabinet	52.
25. Check reaming and rolling gear	53.
26. Hang blinds	54.
27. Blind rack	55.

Continued

Box 6.1: Bapco T&I checklist (c.1995 adapted)—cont'd

Pre-shutdown materials	Other items
1. Assemble column drawings, demister pads, baffles, etc. 2. Identify prefabricated items 3. Draw major materials from stores 4. Label exchanger gaskets 5. Draw column/vessel gaskets from stores 6. Catalyst 7.	1. First shift permits 2. OPD organization 3. Inspection organization 4. 5. 6. 7. 8.

Box 6.1 shows Bahrain Petroleum Company's turnaround and inspection (T&I) checklist used in this phase. The checklist is divided as follows:

A. Shutdown preparation
B. Organization
C. Site Preparation
D. Preshutdown materials
E. Other items.

A final review meeting should take place a few weeks before the turnaround to review the overall preparedness. A key readiness audit also takes place at the start of this phase (see Section 5.8.3 for details).

6.2 Scope

6.2.0 Introduction and relevant PM processes

Phase 4 is the start of execution of the work. Rigorous control of the scope is initiated and all work required to be done prior to commencement of the turnaround must be completed.

The relevant standard processes are:

PMBOK	ISO 21500
5.6 Control scope	4.3.14 Control scope

See Appendix A2 for details.

6.2.1 Required preliminary work

Scope is frozen prior to commencing pre-turnaround work.

Scope for pre-turnaround work must be in accordance with the agreed work plan.

Prefabrication and preparation work is maximized in this phase. Erecting of scaffolding and partial removal of insulation may be permitted in certain areas. For the replacement of fluidic catalytic cracking cyclones, the replacement head with the attached cyclones is placed on a frame ready for installation while a second frame is prepared to receive the old head and cyclones. The crane for lifting these components is also positioned.

All preliminary work is to be completed before feed cut-off ("oil-out" in refining terms) to ensure that, in phase 5, all resources are dedicated only to the work that requires the plant to be out of service.

6.3 Schedule
6.3.0 Introduction and relevant PM processes

Phase 4 calls for the implementation of all on-site pre-turnaround work. Progress is normally reported daily. The pre-turnaround work must include everything possible to reduce unnecessary work during the execution phase of the turnaround. Although progress is monitored, the clock only starts ticking when the flow of product stops.

The relevant standard procedures are:

PMBOK	ISO 21500
6.6 Control schedule	4.3.24 Control schedule

See Appendix A3 for details.

6.3.1 Pre-turnaround execution activities

Pre-turnaround activities need to be maximized to decrease time during the turnaround. Some of these activities are:

- Installation of scaffolding for access
- Insulation removal prior to access
- Lay-down areas for new/replacement equipment ready for lifting into place
- Load-bearing plates over bridges, etc. for heavy vehicle access to areas of the turnaround
- Sand bag oily water sewers
- Preparation for isolation and cleaning
- Tagging isolation points and preparing gaskets and blinds for installation
- Temporary flushing lines, fittings, and equipment
- Tagging of valves to be replaced or repaired
- Lay/build bases for tower cranes and stands for new fluid catalytic cracking unit (FCCU) cyclones

  Examples: Turnarounds for FCC cyclone replacement.

  In the case of an FCCU cyclone replacement in Bahrain Refinery, a base was built for the tower crane to be used for the cyclone replacement. Conversely, for a similar replacement project in

Qatar Petroleum, a rig was built to support the replacement cyclones, which were preattached to the regenerator head in the factory (see Box 4.9).

- Installation of replacement columns or vessels next to the old ones (if space is available)

  Example: Chevron Cape Town Refinery turnaround.

  Before the turnaround, a new de-ethanizer column was erected next to the one that needed to be replaced, trays were installed, the column was hydro tested and insulated, and ladders and walkways were installed.

  During the turnaround, the piping to and from the old column was rerouted to the new one, and the old column and its foundation were removed.

  Extra work involved the building of a new foundation and new piping, but the time saved during the turnaround was significant.

6.3.2 Pre-turnaround work completion

Baseline time, cost, and resourcing is set in the critical path model.

Progress on pre-turnaround activities is monitored and corrective action taken to ensure completion before the start of the turnaround execution phase.

6.4 Cost

6.4.0 Introduction and relevant PM processes

In phase 4 expenditure on pre-turnaround work commences. The cost control model is also tested and contract cost control is initiated, but agreed man-hours for the turnaround execution are excluded.

The relevant standard processes are:

PMBOK	ISO 21500
7.4 Control costs	4.3.27 Control costs

See Appendix A4 for details.

6.4.1 Testing the cost management systems

The critical path model has now been established and frozen, and the baseline has been set to measure progress against the plan (time and cost).

The reporting tools need to be tested, especially the application of the earned value method.

Earned value could be represented as shown in Fig. 6.1.

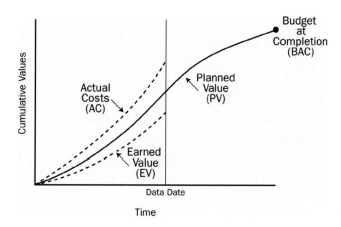

FIGURE 6.1

Earned value graph.

6.5 Quality

6.5.0 Introduction and relevant PM processes

In phase 4 quality control teams are mobilized. Orientation and other related training takes place. Quality control equipment is assembled and tested. (X-ray, hydro-test gauge and pump, ultrasonic machine, infrared cameras, post weld heat treatment rigs, etc.)

Quality control (QC) commences on prefabricated work. In some cases poor quality control at this stage could have a major impact on the turnaround duration. Box 6.10 discusses the necessity to maximize prefabrication.

The relevant project standard processes are:

PMBOK	ISO 21500
8.2a Manage quality	4.3.33 Perform quality assurance
8.3a Control quality	4.3.34 Perform quality control

See Appendix A5 for details.

6.5.1 Pre-turnaround quality activities

Quality control plans are issued with work packages.

QC carried out on prefabrication work is used to test the effectiveness of the QC processes and procedures.

6.5.2 Resourcing for quality control

Adequate numbers of qualified inspectors certified for X-ray interpretation, etc. need to be obtained. A contract with an inspection authority may be required, or, alternatively, the company's affiliates may be in a position to offer inspectors.

Example

Chevron sent inspectors from their affiliates to support the inspection activities of turnarounds at other affiliates.

The level of resourcing will be determined by the critical path model as all inspection activities need to be totally integrated into work activities.

6.5.3 Receipt of materials on-site

Materials delivered to the site for a turnaround need the required certification. On-site verification is needed for composition of alloys. Box 6.2 outlines the uses of positive material identification (PMI) test guns.

6.6 Resources

6.6.0 Introduction and relevant PM processes

Resourcing is ramped up in anticipation of shutting down the plant. Intensive orientation, training, and team building exercises take place for the staff and contractors who have been mobilized. Logistics for security clearance, accommodation, and locating equipment and materials, prior to the start of the turnaround, are set into motion.

Box 6.2: Positive material identification (PMI) test guns[1]

Uses include:
- PMI for Alloys
- PMI for Scrap Metal
- Alloy Grade Verification
- PMI for Metal Fabricators
- PMI for Particular Metal Alloy Groups:
  - Steel
  - Stainless Steel
  - Carbon and Low-Alloy Steel
  - Copper Alloys, Bronze, Brass
  - Aluminum Alloys
  - Titanium Alloys
  - Nickel, Cobalt, High-Temperature Alloys
- Weld Verification
- Component Composition Verification
- PMI for High-Reliability ("Hi-Rel") and Exempt Alloys
- PMI for Petrochemical Refinery Piping and Components.

Relevant standard processes are:

PMBOK	ISO 21500
9.3b Acquire resources (contractors)	–
9.4c Develop team	4.3.18 Develop project team
9.5a Manage team	4.3.20 Manage project team
9.6a Control resources	4.3.19 Control resources

See Appendix A6 for details.

6.6.1 Team training

Team training now encompasses a much larger team that includes all of the contractor supervisors as well as others requiring specific skills for the shutdown.

Technical team members require training in the following:

1. Shutdown scope
2. Technical team member roles and responsibilities
3. Contingency plans
4. Extra work flow and paper flow
5. Shutdown organization.

Technical team members also need to be appraised of the following as part of their training:

1. Available support resources such as operations, inspection, engineering, maintenance, purchasing, clerical, etc.
2. Support resource working hours, 24 h phone/pager numbers, and if contactable by radio (published list).

Contractor supervisors require training in at least the following:

1. Plant orientation
2. Safety
3. Permit receipt.

Dedicated **contract workers** require training in these areas:

1. *All*—plant orientation
2. *Fire watcher*—where hot work is carried out
3. *Hole watcher*—at points of access to pressure vessels and furnaces
4. *Planner*—introduction to the planning model established specifically for the upcoming shutdown with special reference to standardized work breakdown structures, organization breakdown structures, calendars, and resource codes.

Shift operating staff require training as follows:

1. Fire watch—where hot work is carried out
2. Hole watch—at points of access to pressure vessels and furnaces
3. Specific assignments with respect to preparation of the plant for shutdown and start-up
4. New operating procedures for plant modifications as required by the management of change (MOC) process.

Maintenance staff require training as follows:

1. Installation and maintenance of new equipment as required by the MOC process.

A certain number of specific staff such as **supervisors and administrators** require the following:

2. **First aid**—the standard basic first aid course to ensure there is the correct number of first aiders for the number of contract staff, with first aid certificates that are current
3. **Fire Fighting**—the basic fire-fighting course to ensure that there are sufficient numbers of trained fire fighters with certificates that are currently valid.

As the shutdown and start-up of process units are the most dangerous times, an **emergency response training exercise** should have been carried out within 6 months prior to the shutdown.

6.6.2 Team building

To build a spirit of cooperation between contractors, and between staff and contractors, a social get-together of staff and contractor supervisors prior to the turnaround is recommended.

> *Example*
>
> *Common practice in South Africa is to have a pre-turnaround social "braai" (barbeque) to which all contractor and client supervision are invited. The intent is to introduce the members of the leadership group to each other in a relaxed social environment.*
>
> *Contractors also sign an "agreement of cooperation" linked to the "turnaround philosophy" (see Section 3.1.2).*

6.6.3 Equipment and tools

Equipment and tools start to arrive on-site and need to be located according to the plot plan.

6.6.4 Receipt of major equipment on-site

Further to Section 5.9.5 where the receipt of major equipment is planned, resources are now required to get the equipment to site and, in some cases, to erect it. Logistics plays a major role requiring supporting man-power, equipment, and materials. Box 5.14 details an example.

6.6.5 Resourcing the safety team

The safety team would consist of full-time and part-time members drawn from both the client and the contractor. Members include:

1. Client safety officer (full-time)
2. Contractor safety officers (full-time from each major contractor)
3. Safety representatives (part-time client and contractor representatives)
4. Turnaround emergency controller (part-time client representative)
5. Emergency marshals (part-time selected from client/contractor supervisors)
6. Confined space entry guardian (full-time selected from operators that are available or contractors).

6.7 Communication

6.7.0 Introduction and relevant PM processes

In phase 4 communication intensifies after freezing of the schedule and mobilization of the contractors. Briefings and notifications are used to communicate with those actioners (implementers) who have not been involved with the planning of the turnaround. HSE communication is of prime importance to ensure an incident-free turnaround. Required documentation is distributed.

Related standard processes are:

PMBOK	ISO 21500
10.2a Manage communications	4.3.40 Manage communications
10.3a Monitor communications	4.3.39 Distribute information

See Appendix A7 for details.

6.7.1 Mobilization

The logistics planned in Section 4.7.1 are implemented.

Contracts are awarded and the key contractor staff start to arrive on-site for orientation and to set up the site offices. Administration facilities, computers, and phones are installed in mobile offices close to the activities as designated by the client. Lists of staff for security clearance and verification of qualifications are submitted to the client. Mobile equipment is brought on-site.

Plant preparation jobs are carried out and include:

1. Tagging for blinding, valve repair or replacement, removal of piping, etc.
2. Pallets of blinds are located at required areas throughout to plant.
3. Scaffolding is erected where it cannot interfere with normal operation of the plant.
4. Oily water sewer drains are sandbagged to prevent gas emissions.

5. Additional area lighting is installed for night activities.
6. Temporary connections for flushing, chemical cleaning, etc. are prepared.
7. Materials are drawn from the store to area holding pens close to the process.

"Job walk-through" is required for key field supervisors and engineers with drawings specs, etc.

6.7.2 Final execution plan

Once the **final budget decision (FBD)** is made, final tuning of the planning model is carried out up until **the scope freeze date**, which is typically about 6 weeks before the shutdown. All change requests after the **FBD** must go through the change control process.

On the scope freeze date, the planning model critical path and resource leveling is done for the final time before the **turnaround baseline** is set. This is the basis for measuring progress and expenditure variance.

The first official issue of all required reports is then distributed. Examples of reports are:

General viewing in the shutdown activity center

1. Organization breakdown structure
2. Skeleton network (product out to product in)
3. Summary Gantt charts by process units
4. Four-day look ahead for near critical and critical activities for each unit
5. Earned value curves
6. Resource histogram (overall)
7. Productivity curves

Top management

1. Earned value curves—most valuable for seeing the big picture
2. Summary Gantt charts by process unit
3. Four-day look ahead for near critical and critical activities for each unit
4. Exception reports: listing activities behind schedule, cost overruns, additional work and costs, quality excursions, HSE incidents.

Area supervisors

1. Four-day look ahead for all work for which they are responsible.
2. Construction work packages including drawings, specifications and quality control procedures (QCPs).

Should the correct codes have been entered when the planning model was created, nearly any type of report could be produced on demand.

Once the shutdown commences, **daily** processing of data and reporting should be carried out.

Box 6.3 shows a skeleton report for a complete refinery turnaround at the start of the turnaround.

Box 6.3: Master skeleton turnaround plan published just prior to a turnaround

Summary of all process units to be shutdown

Note:

No 2 FCC and platformer process units are on the critical path (zero total float).

ID	Activity Desc.	Dur	TF	Early Start	Early Finish
005	NO 2 FCCU	1041h	0	15/04/96 06:00	28/05/96 15:00
035C	PI 63 HOT WORK	32d	1d	19/04/96 07:00	21/05/96 07:00
040	CAT POLY PLANT	597h	444h	15/04/96 06:00	10/05/96 03:00
067	PI 70 HOT WORK	500h	444h	19/04/96 07:00	10/05/96 03:00
070	ISOM PLANT	20d	593h	14/04/96 16:00	04/05/96 16:00
077	PI 71 HOT WORK	18d	593h	16/04/96 16:00	04/05/96 16:00
080	SRU Train No2	739h	201h	13/04/96 12:00	14/05/96 07:00
116	PI 67 (SRU No 2) HOT WORK	25d	362h	19/04/96 07:00	14/05/96 07:00
120	PLANT 55	473h	466h	20/04/96 06:00	09/05/96 23:00
147	PI 55 HOT WORK	400h	466h	23/04/96 07:00	09/05/96 23:00
	KERO HYDROTREATER	1065h	91h	15/04/96 00:00	29/05/96 09:00
170A	PI 6 Second HOT WORK period	401h	363h	22/04/96 21:00	09/05/96 14:00
175	DIESEL HYDROTREATER	41d	167h	15/04/96 00:00	25/05/96 24:00
195A	PI 56 Second HOT WORK period	479h	189h	22/04/96 21:00	12/05/96 20:00
200	NO 1 CDU	973h	132h	14/04/96 18:00	25/05/96 07:00
230	PI 2 Second Hot Work period	508h	230h	22/04/96 21:00	14/05/96 01:00
235	NO 1 VOU	973h	132h	14/04/96 18:00	25/05/96 07:00
242	PI 52 Second Hot Work period	531h	225h	22/04/96 21:00	14/05/96 24:00
270	NAPTHA HYDROTREATER	982h	31h	15/04/96 06:00	26/05/96 04:00
296	PI 3 Second Hot Work period	637h	31h	22/04/96 21:00	19/05/96 10:00
297	DHT DIST SECT	52h	167h	15/04/96 02:00	17/04/96 06:00
300	NO 2 CDU	965h	86h	15/04/96 06:00	25/05/96 11:00
330	PI 60 HOT WORK	626h	195h	19/04/96 06:00	15/05/96 08:00
335	NO 2 VDU	973h	86h	15/04/96 06:00	25/05/96 19:00
365	PI 61 HOT WORK	753h	86h	19/04/96 06:00	20/05/96 15:00
	AMINE DEA	32h	151h	15/04/96 06:00	16/04/96 14:00
389A	PLATFORMER	924h	0	15/04/96 12:00	23/05/96 24:00
392	PI 4 Second Hot Work period	501h	9h	22/04/96 21:00	13/05/96 18:00
393	Platformer DIST SECT	897h	31h	15/04/96 18:00	23/05/96 03:00
394	Second Hot Work period	649h	31h	22/04/96 21:00	19/05/96 22:00
395	VISBREAKER UNIT	43d	110h	14/04/96 18:00	27/05/96 18:00
425A	PI 5 HOT WORK	31d	39h	23/04/96 02:00	24/05/96 02:00
430	NO 1 FCCU	873h	22h	20/04/96 06:00	26/05/96 15:00
461	PI 53 HOT WORK	500h	22h	23/04/96 07:00	14/05/96 03:00
461AA	TANK 507	596h	236h	23/04/96 07:00	18/05/96 03:00
461BA	PL 20 HOTWORK	428h	236h	29/04/96 07:00	17/05/96 03:00
470	SRU Train No 1	573h	56h	20/04/96 06:00	14/05/96 03:00
500A	PI 67 HOT WORK	500h	366h	23/04/96 07:00	14/05/96 03:00
509	FLARE SYSTEM	18d	310h	22/04/96 07:00	10/05/96 07:00
510A	PI 101 HOT WORK	16d	310h	23/04/96 07:00	09/05/96 07:00
519	STEAM SYSTEM	29d	22h	23/04/96 07:00	22/05/96 07:00
519A	DEAERATOR REPAIRS	24d	22h	23/04/96 07:00	17/05/96 07:00

Apr bar chart columns: 13 14 15 16 17 18 19 20 21 22 23 24 25 26 27 28 29 30 31

Box 6.3: Master skeleton turnaround plan published just prior to a turnaround—cont'd

No 2 FCC shutdown and start-up activities opened.

Notes:

Detailed FCC network models are below activities 035AA, 035B and 035C.

Red summary bar 035AA/ has critical path activities in the detailed network model below it.

ID	Activity Desc.	Dur	TF	Early Start	Early Finish	Apr
						13 14 15 16 17 18 19 20 21 22 23 24 25 26 27 2
001	START 1996 TURNAROUND (time ⌐	0	0	13/04/96 12:00	13/04/96 12:00	
005	NO 2 FCCU	1041h	0	15/04/96 06:00	28/05/96 15:00	
010	shutdown	8h	0	15/04/96 06:00	15/04/96 14:00	
015	swing blinds	4h	0	15/04/96 14:00	15/04/96 18:00	
020	water wash	13h	0	15/04/96 18:00	16/04/96 07:00	
025	steaming	24h	0	16/04/96 07:00	17/04/96 07:00	
030	blinding	24h	0	17/04/96 07:00	18/04/96 07:00	
035	resteam	16h	0	18/04/96 07:00	18/04/96 23:00	
035A	Open manways	8h	0	18/04/96 23:00	19/04/96 07:00	
035AA	PI 63 FCCU T&I	768h	0	19/04/96 07:00	21/05/96 07:00	
035B	PI 63 GCU T&I	768h	1d	19/04/96 07:00	21/05/96 07:00	
035C	PI 63 HOT WORK	32d	1d	19/04/96 07:00	21/05/96 07:00	
036	Box up structure	24h	0	21/05/96 07:00	22/05/96 07:00	
036A	Check out	48h	0	22/05/96 07:00	24/05/96 07:00	
036B	MBA Steam out	24h	0	24/05/96 07:00	25/05/96 07:00	
036C	Fuel gas	24h	0	25/05/96 07:00	26/05/96 07:00	
036D	Fire up Peabody	24h	0	26/05/96 07:00	27/05/96 07:00	
036E	Start Cat Circulation	32h	0	27/05/96 07:00	28/05/96 15:00	
036F	Feed In	18h	0	28/05/96 15:00	29/05/96 09:00	

Box 6.4 shows different Gantt charts at the start of a turnaround.

Box 6.5 shows a typical resource histogram.

6.7.3 HSE relationships

Behavior-based safety is critical for a successful incident-free turnaround. Fig. 6.2 shows the primary relationships for a turnaround.

6.7.4 Briefings and notifications

Briefings encourage questions and answers, whereas notifications are one-way communications. Notifications are conveyed through newsletters and on bulletin boards. The use of an internal company newsletter is useful to communicate major issues such as planned major work and HSE enforcement on-site during the turnaround.

Box 6.4: Sample Gantt reports[2]

Summary Gantt Chart: Overview of a complex.

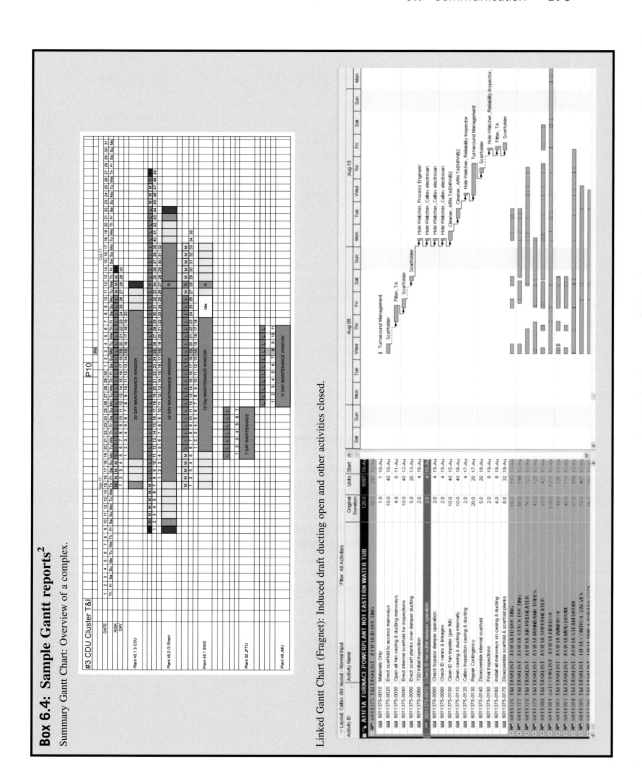

Linked Gantt Chart (Fragnet): Induced draft ducting open and other activities closed.

Box 6.5: Sample resource histogram

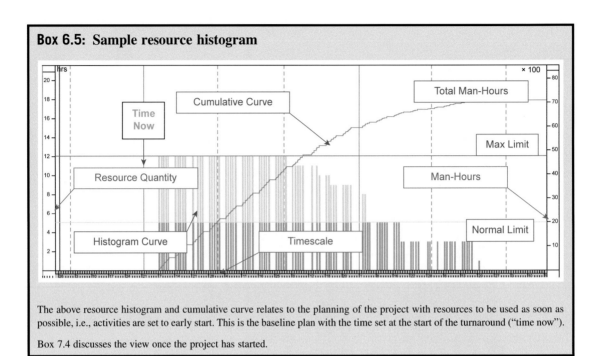

The above resource histogram and cumulative curve relates to the planning of the project with resources to be used as soon as possible, i.e., activities are set to early start. This is the baseline plan with the time set at the start of the turnaround ("time now").

Box 7.4 discusses the view once the project has started.

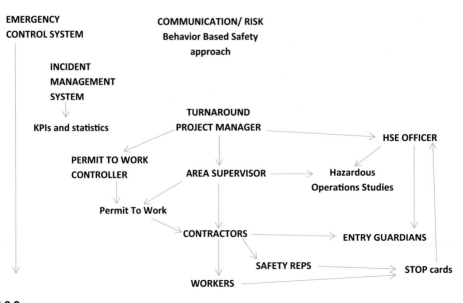

FIGURE 6.2

Health, safety, and environment (HSE) relationships.

General briefings

Since a turnaround necessitates a large amount of work carried out in a short time by many people, it is prudent to have a series of briefings to ensure alignment with the turnaround philosophy. Elements include:

1. Purpose and objectives of the turnaround
2. Turnaround organization
3. Key dates including shutdown and start-up dates
4. Working patterns including different shifts and trades involved
5. Scope of the turnaround in brief, as well as a strictly enforced scope change process
6. Contractors and their specialization expertise
7. Cost plan
8. Quality requirements
9. HSE issues including a streamlined permit to work process and hole watch resourcing
10. Facilities including messing, changing, toilets, etc.

The turnaround project manager is to lead the presentation and discussion, with the sponsor/decision executive showing active visual support where appropriate.

Notification of major work

An example of communicating major maintenance work in a turnaround is described in Box 6.6.

Box 6.6: Communication example—major work

Internal Refinery News Article—FCC repairs

Changes to No. 1 FCCU.

During the upcoming T&I, major changes will be taking place on the No. 1 FCCU. The riser, reactor vessel, regenerator cyclones, regenerator air grid, and orifice chamber will all be replaced. Replacement of these components is necessary since they reached the end of their productive lives.

The existing curved riser will be replaced with a vertical riser, micro-jet feed injection nozzles and direct-coupled cyclones, similar to the equipment installed in the No. 2 FCCU during the last T&I. ABB Lummus in Houston, Texas, performed the process engineering for the new riser and other components in accordance with our specifications. Lummus supplied the latest FCCU technology that will result in improved yields from the unit. Replacing the regenerator cyclones will improve the efficiency of catalyst removal from the flue gas, thus reducing emissions to the atmosphere.

Foster Wheeler SA performed the detailed design and procurement for the project, based on the basic engineering completed by Lummus. During SD2K Foster Wheeler field engineers will be on-site to provide technical support.

Construction work will be completed by our alliance construction contractor, CB&I. Currently CB&I is installing the structural steel and the lower sections of the new riser on the north side of the existing FCCU structure. Prior to the T&I we hope to have the lower sections of the riser, the new regenerated catalyst withdrawal well, and associated pipework installed. The new regenerator cyclones are presently being assembled in the asphalt plant; if you have some spare time go and look at them, it is not often that you will have the an opportunity to view a set of cyclones up close.

During the shutdown the big changes will take place. The old reactor will be lifted out in two pieces and the new reactor, weighing approximately 80 tons, will be lifted into position in one piece. Once the new reactor is installed, the new riser and structural steel will be completed. Removal of the old regenerator cyclones will be accomplished by cutting the top head of the regenerator, using high-velocity water jets. The head and existing cyclones will then be lowered to the ground where the new cyclones will then be fitted to the head before replacing it. To facilitate replacement of the regenerator air grid and refractory, a section of the regenerator bottom head will also be removed.

Health, safety, and environment awareness notification

HSE awareness is an essential part of managing the turnaround to ensure zero major incidents, a turnaround key performance indicator. An example of communicating HSE requirements during the turnaround is given in Box 6.7.

Public notification

External notifications inform the general public about what is happening on-site. An example is shown in Box 6.8.

Box 6.7: Communication example—HSE

Internal Refinery News Article—HSE

The correct (and essential) use of personal protective equipment (PPE)—before and during the shutdown
Protective clothing must be provided, free of charge, to all employees by their employers and it is mandatory to wear the equipment received.

Safety glasses
The refinery has a policy on the wearing of safety glasses, which requires all personnel on site (Caltex staff and contractors) to wear safety glasses at all times, except inside buildings. Please note, however, that the general-purpose safety glasses do not protect eyes against ultraviolet or infrared rays arising from welding or grinding operations. Special safety glasses must be used during such work.

A high rate of eye injuries have occurred in past major shutdowns, to personnel working close to welding or grinding operations without adequate protection.

Hard hats
The wearing of hard hats is mandatory on all roadways and in process plant and tank farm areas. All contractors have been requested to provide hard hats for their staff and must ensure that their company logo is clearly identifiable on each employee's hard hat.

Only people who are en route to work having just arrived on site, or are leaving site, are excused from wearing hard hats.

Ear plugs
It is accepted that once the plant is shut down, the nonshutdown demarcation of areas as noise zones may no longer be valid. Nevertheless, machinery for shutdown use will be operational throughout the refinery.

There is a simple test to determine if the noise in the work area is sufficient to warrant the use of ear protection. If a person speaking to you needs to raise their voice when they are within 1 m of you, then the noise level is above 85 dB(A), and, therefore, requires the use of ear plugs/muffs.

Overalls
All overalls must have long sleeves and be closed at the front. Loose-flapping clothing can be caught in machinery, and unprotected arms can be burnt against steam lines and hot process piping.

Safety footwear
Safety shoes or boots must be worn in workshops, process areas, and off-site facilities. Footwear must be nonslip, oil resistant, and have steel-toe caps.

Respirators
An issue store is available at the main stores for personnel requiring respirators. The store provides all the respiratory equipment required (disposable respirators, half or full face masks with cartridges, and air-supplied respirators).

Used equipment must be returned to this store for cleaning and maintenance immediately after use.

A person competent in respirators is available to advise and issue workers with the correct type of respirator required for the various jobs. Supervisors are requested to either collect the respirators from the store and/or inform workers, who are sent to collect their own equipment, of the hazards associated with the job.

Self-contained breathing apparatus will be issued to safety watchers from the fire department store.

> **Box 6.8: Turnaround press release example**
>
> **Neste Oil's Porvoo refinery prepares for major maintenance turnaround**
> Neste Oil Corporation.
>
> Press Release (edited).
>
> 17 August 2005 at 2:00 p.m. EET.
>
> Neste Oil will carry out a planned five week maintenance turnaround at its Porvoo refinery site beginning 22 August when shutting down the refinery's units will begin. The actual maintenance will start on 29 August, and the refinery should be back in normal operation by the end of September.
>
> The turnaround which is part of the normal maintenance and modernization program at Porvoo will enable the refinery to retain its high capacity utilization potential over the next five years. Various statutory pressure vessel inspections will also be carried out, together with other work to ensure it remains a low environmental impact facility....
>
> *Safety will be prioritized*
> Safety is a major priority for Neste Oil during shutdowns, start-ups, maintenance work, and capital projects. Safety training of company personnel for the upcoming turnaround was started in the spring. The employees of contractors that will be involved have received comprehensive safety training for the project, emphasizing safety thinking, following safety instructions, and careful risk assessment on the job.
>
> A total of around 2300 employed by various contractors will be on site at Porvoo during the turnaround. The turnaround will employ some 3100 people.
>
> The 5 week turnaround and associated loss of production will adversely impact Neste Oil's profit for the third quarter and for 2005 as a whole. In its second-quarter interim report the Company estimates losses to be EUR 40–60 million on the Group's operating profit.
>
> The previous major maintenance turnaround at Porvoo took place in spring 2001.
>
> *Adapted from Neste Oil Corporation Press Release 17 August 2005[3].*

6.8 Risk (including HSE issues)
6.8.0 Introduction and relevant PM processes

In phase 4 everything feasible is done to mitigate risks during the execution of the turnaround. Completion of all possible work prior to the turnaround is essential to ensure no unnecessary preparation work is carried out in the turnaround execution phase. Testing of systems and processes to reduce risk of deviation from procedures is also needed. Emergent risks are to be addressed promptly. The risk register is updated to keep it current.

Related standard processes are:

PMBOK	ISO 21500	ISO 31000 risk management
11.2 Identify risks	4.3.28 Identify risks	1. Establish the context
—	—	2. Communicate and consult
11.3 Perform qualitative risk analysis	4.3.29 Assess risks	3. Risk assessment
11.4 Perform quantitative risk analysis		Risk identification
		Risk analysis
		Risk evaluation
11.5 Plan risk response	4.3.30 Treat risks	4. Risk treatment
11.6 Implement risk responses		
11.7 Monitor risks	4.3.31 Control risks	5. Monitor and review

See Appendix A8 for details.

6.8.1 Safety, health, environment and quality control

An incident management procedure for safety, health, environment, and quality (SHEQ) is a requirement in the oil and gas industry. This is required to log major reportable incidents for one of the key performance indicators.

Other procedures, guidelines, and checklists need to be implemented, these being a few examples:

1. Permit issuing procedure
2. Confined space access procedure
3. Confined space air testing procedure
4. Unit cleaning and preparation procedure
5. Lifting procedure
6. Scaffolding procedure
7. Shutdown environmental guidelines
8. Blind installation and removal procedure
9. Gasket replacement procedure
10. Orifice plate removal and replacement
11. Valve repair and replacement procedure
12. Equipment checklist
13. Pressure test certificate
14. Column and vessel man-way checklist
15. Welders' records
16. Welding procedure
17. Product quality procedures for plant start-up.

A worthwhile memory jog regarding potential complications would be a quick read of Trevor Kletz's book *What Went Wrong*, especially Chapter 1.[4]

6.8.2 Change control

Unauthorized changes pose a serious risk to completing the turnaround on time and on budget.

The MOC process is discussed further in Section 7.1.4.

There are *two* types of change, in-kind and not-in-kind:

- **In-kind** change is an identical replacement that requires no redesign, documentation, or additional training.
- **Not-in-kind** change is any modification to equipment and procedures.
  A number of **not-in-kind** changes take place during turnarounds.

   Examples

   1. *Replacement or addition of equipment other than in-kind*

> **Box 6.9: Flixborough disaster[5]**
>
> "The explosion at the Flixborough Nypro Chemicals site near Scunthorpe, UK, killed 28 people and injured 36 others on 1 June 1974. It resulted in the almost complete destruction of the plant. Further afield, the blast injured another 53 people and caused extensive damage to around 2000 buildings."
>
> "The cause of the disaster was the installation of a **temporary pipe** to allow production to continue whilst a large reactor was removed for repair. The failure of a flexible bellows next to the temporary pipe, led to a massive release of boiling *cyclohexane* which ignited causing the explosion."

2. *Operating outside safe, established limits as could occur in shutdown and start-up of the process*
3. *Use of new chemicals, feeds or catalysts*
4. *Modification or addition to operating, health and safety, or maintenance procedures*
5. *Use of alternative construction materials*
6. *Temporary connections and repairs.*

Box 6.9 gives an example of what can happen without a structured MOC process. Subsequent regulations in Europe enforced, among other things, the MOC process.

It is critical that a formal MOC process is adhered to, and a procedure needs to be in place to manage these changes. This is a critical time for reviewing identified changes to ensure that necessary specifications are adhered to, and that the required training of operating and maintenance staff is carried out.

6.8.3 Hazardous substances

Identification of hazardous substances is discussed in Section 4.8.3. Removal of hazardous substances and preparation for cleaning internals in anticipation of entry thus requires resourcing.

Examples

1) *Substances such as benzene (carcinogenic) in process internals. Flushing and disposal of contaminated fluids is required.*
2) *Heavy metals such as vanadium in furnaces, boilers, and flues. Wash down of surfaces and disposal of wash water is required.*
3) *Asbestos in cladding. Removal, bagging, and disposal are required.*

Typical asbestos removal includes:

1. Identifying the area for asbestos removal
2. Sealing off the area completely
3. Workers donning appropriate personal protection equipment (including breathing apparatus)
4. Removing, bagging, and sealing bags of asbestos inside the sealed area
5. Removing the bags to an approved disposal site.

6.8.4 Audit

A final pre-turnaround audit is required to ensure that everything possible has been completed in anticipation of the execution phase of the turnaround.

6.9 Procurement

6.9.0 Introduction and relevant PM processes

In phase 4, contractors are mobilized and execution of the turnaround contracts commences. Pre-turnaround contractor training and testing takes place and all on-site prefabrication is completed. The receipt of materials to site is closely monitored (see Section 6.5.3).

Related standard processes are:

PMBOK	ISO 21500
12.3 Control procurements	4.3.37 Administer procurements

See Appendix A9 for details.

6.9.1 Prefabrication

Maximum prefabrication is required to minimize downtime, and construction methods have to be critically reviewed to achieve this.

Where possible, modular prefabrication should take place. For example, the building of foundations and the placing of modular units may be permitted prior to the start of the shutdown.

Whether the prefabrication is done on- or off-site, a thorough investigation of sizes of prefabricated components, transport modes, and routes to the process unit is needed at the design stage.

Prefabrication progress should be monitored using critical path and cost control on the planning model.

Box 6.10 explains the need for major prefabrication in certain cases.

6.9.2 Contractor mobilization

Security clearance and qualification verification/validation is carried out. The contractors are then mobilized as per the requirements of the critical path method model. Mobilization is discussed in detail in Section 6.7.1.

Training and orientation are discussed in Sections 6.6.1 and 6.6.2.

Box 6.10: Post-weld heat treatment

A large compressor suction line was required to be replaced in a turnaround. This requires extensive welding where every weld was to be post-weld heat treated (PWHT). This takes time especially if each weld is done individually. It was decided to prefabricate the suction line in two pieces since the factory had an oven large enough to accommodate them for PWHT. This resulted in only three welds on-site with associated PWHT. Turnaround time for this work was vastly shortened and quality control was much enhanced with it being done mostly before the execution phase in a factory environment.

In another case, extensive post-weld heat treatment took place during the turnaround, which became a primary factor in extending the turnaround by 2 weeks. This could have been preempted by maximizing prefabrication as in the above case (see Box 5.15).

6.10 Stakeholder management
6.10.0 Introduction and relevant PM processes

Final approval would have been secured by the end of phase 3. In phase 4 stakeholder awareness with respect to what is going to be carried out is completed and notification is given to the key stakeholders, including:

1. Government health, safety, and environment authorities
2. Suppliers of raw materials with requests to suspend supply
3. Customers to notify them of possible changes of product delivery from stockpile or other sources.

Related standard processes are:

PMBOK	ISO 21500
13.3 Manage stakeholder engagement 13.4 Monitor stakeholder engagement	4.3.10 Manage stakeholders

See Appendix A10 for details.

6.10.1 Final approval

Decision-maker final approval (go/no go) is required to commence the turnaround based on the final front-end loading audit, including the availability of all materials required for the turnaround and the readiness of the contractors to mobilize.

On occasion, turnarounds have been postponed by weeks, or even months, as key materials are not on-site or a key contractor is having difficulty mobilizing. Alternatively, product buildup may be insufficient, requiring a few more days of production.

> *Example*[6]
>
> "Ecuador postponed planned maintenance on its 110,000 barrel-per-day Esmeraldas refinery due to a delay in the delivery of pipes ...." *The delay was about 7 months.*

6.10.2 Government and other regulatory authorities
Health and safety

Notification of inspection on boiler and other pressure vessels is required in terms of the law or insurance policy.

Any major incident (including loss of life) needs to be reported to the relevant authorities.

Environment

Gas plants may be required to report the excess flaring that occurs when a gas plant is shut down or started up to a regulatory authority.

Power and water

Major changes in power demand requires notification to the electricity supply authority.

Excess use of water for flushing and cleaning may require notification to the water supplier and regulatory authority.

6.11 Summary: phase 4—pre-execution work
6.11.1 Focus areas

1. SHEQ plan implemented
2. Training for contractors—orientation, fire watch, hole watch, permit receipt, etc.
3. Training for operators—fire watch, hole watch, permit issue, etc.
4. MOC requirements review
5. Team building
6. Turnaround sequences detailed
7. Plant walk-through
8. Temporary connections for sweetening, etc.
9. Tagging—blinds, cut-ins, etc.
10. Contractor mobilization
11. Cost tracking and reporting.

6.11.2 Deliverables

1. Execution plan finalized (frozen) and published
2. Organization chart published
3. Preexecution work completed
4. Field mobilization started
5. Audit report recommendations completed
6. Training completed.

6.11.3 Tools and processes

1. Turnaround execution guideline/procedure
2. Planning model
3. Audit function

6.11.4 Timing

From 3 months to 2 weeks prior to the final execution of the turnaround.

References

1. Bruker. Available from: https://www.bruker.com/products/x-ray-diffraction-and-elemental-analysis/handheld-xrf/applications/pmi.html.
2. McGrath R. *The recipe for success: investing in comprehensive shutdown planning STO Asia.* IQPC; 2014.
3. Neste Oil Corporation Press Release 17 August 2005. Available from: www.nesteoil.com.
4. Kletz T. *What went wrong.* 4th ed. Elsevier; 1999.
5. Institution of Chemical Engineers (IChemE). Available from: https://www.icheme.org/media_centre/news/2014/lasting-safety-lessons-as-flixborough-remembered.aspx.
6. Valencia A. *Ecuador postpones Esmeraldas refinery maintenance.* Hydrocarbon Processing; 2018.

Phase 5 Turnaround execution

7.0 Outline

We may squeeze 800,000 man-hours of work into 30 days.[1]

The execution of the turnaround commences with stopping the output of product ("feed out") and ends with product flowing to the customer again at the required quality standard ("feed in"). Activities are discussed under the following headings:

1. integration
2. scope
3. schedule
4. cost
5. quality
6. resources
7. communication
8. risk (including health, safety, and environment [HSE] issues)
9. procurement
10. stakeholder management

Specifically Table 7.1 references PMBOK and ISO 21500 processes applicable in this phase.

7.1 Integration
7.1.0 Introduction and relevant project management (PM) processes

This phase entails intensive management and control of the execution activities to ensure completion of the turnaround on schedule and within budget.

Turnaround Management for the Oil, Gas, and Process Industries. https://doi.org/10.1016/B978-0-12-817454-8.00007-1

The relevant standard processes are:

PMBOK	ISO 21500
4.3b Direct and manage project work	4.3.4 Direct project work
4.4 Monitor and control project work	4.3.5 Control project work
4.5 Perform integrated change control	4.3.6 Control changes

Table 7.1 Phase 5—turnaround execution.

No.	PMBOK	ISO 21500
1	4.3a Direct and manage project work	4.3.4 Direct project work
	4.4 Monitor and control project work	4.3.5 Control project work
	4.5 Perform integrated change control	4.3.6 Control changes
2	5.5b Validate scope (new/additional work)	–
	5.6 Control scope	4.3.14 Control scope
3	6.7 Control schedule	4.3.24 Control schedule
4	7.4 Control costs	4.3.27 Control costs
5	8.2a Manage quality	4.3.33 Perform QA
	8.3 Control quality	4.3.34 Perform QC
6	9.5a Manage team	4.3.20 Manage project team
	9.6a Control resources	4.3.19 Control resources
7	10.2a Manage communication	4.3.40 Manage communications
	10.3a Monitor communications	4.3.39 Distribute information
8	11.2 Identify risks	4.3.28 Identify risks
	11.3 Perform qualitative risk analysis	4.3.29 Assess risks
	11.4 Perform quantitative risk analysis	
	11.5 Plan risk responses	4.3.30 Treat risks
	11.6 Implement risk response	4.3.31 Control risks
	11.7 Monitor risks	
9	12.3 Control procurements	4.3.37 Administer procurement
10	13.3b Manage stakeholder engagement	4.3.10 Manage stakeholders
	13.4 Monitor stakeholder engagement	–

See Appendix A1 for details.

7.1.1 Overview

Turnaround execution is phase 5 of the implementation process. This phase includes process feed reduction, the preparation of the plant for access, the actual turnaround work, handover, and start-up of the plant again.

Daily monitoring of the schedule and cost, corrective action, and control of scope are critical throughout the turnaround.

On completion of the turnaround work, the decision executive/operations manager reclaims the plant and the start-up commences. Execution is over once the feed has been introduced, but continued coverage by maintenance and technical manpower is needed until the plant is stable.

Release to the operations department is the deliverable of this phase.

7.1.2 Execution process

The execution process generally follows this sequence:

1. Shutting down: removing process fluids, decontamination, cooling, isolating
2. Opening up: physical disconnection, access cover/man-way removal
3. Inspecting: visual with or without inspection tools, reporting
4. Overhauling existing items, installing new items, and removing redundant items
5. Boxing up: final internal inspection, replacing covers/man-ways
6. Testing: pressure systems, trip and alarm systems
7. Start-up: reconnecting services and reintroducing process fluids
8. Plant cleanup and inspection.

7.1.3 Performance measurement

Progress is defined as the rate of achievement measured against the plan, to complete the turnaround within the required duration.

The rate of progress is determined by asking the following questions:

A. How far is the turnaround ahead or behind the critical path? (total float in hours)
B. What are the restraints preventing the achievement of the baseline critical path? (if any)
C. Is approved emergent work impacting the baseline critical path?
D. If the turnaround is behind schedule, what measures are being taken to get back on track?

Specific issues resulting in the turnaround being behind schedule include:

- Poor productivity
- Delays in material deliveries
- Shortage of qualified persons

- Rework and misfit problems
- Permit delays
- Schedule area conflicts between contractors
- Excessive work scope changes.

Key performance indicators (KPIs) will identify deviations but not the causes of the deviations.

The following should be tracked:

1. Schedule
   a. Behind/ahead of critical path: hours
   b. **Schedule variance** (SV): actual/planned duration
   c. Schedule performance index—alternative to SV
      i. Deviation from planned to date: earned value/planned value
2. Cost
   a. **Cost variance** (CV): actual/planned costs
   b. Cost performance index—alternative to CV
      i. Deviation from budget to date: earned value/actual cost
      ii. Deviation from budget: estimate at completion/budget at completion
      iii. Additional work: Actual versus contingency %
3. Scope
   a. Emergent work
      i. Man-hours as % of total turnaround man-hours: % *(minimize)*
4. HSE
   a. Lost time incidents: number *(target: zero)*
      i. Monitor lower-level HSE indicators to prevent even one incident
   b. Flawless start-up (no incidents, **no leaks**): number *(target: zero)*
      i. Measure from "hand back" to "on-stream at required quality"
5. Productivity
   a. Productivity Index
      i. Actual total man-hours on the job/total clocked man-hours.

One to three are derived from the critical path model on a daily basis, whereas four and five are derived from other sources.

7.1.4 Scope change control

Once the final work list has been approved, all changes must go through a strict change control procedure, and all changes must have unique change request numbers for tracking.

A key element of this process is the effect that the change could have on the shutdown duration and cost. The change is simulated in a copy of the shutdown model to gauge the potential influence on the turnaround, and the result determines whether it should be included or not.

Note that a scope change may, or may not, entail the management of change (MOC) process (see Section 6.8.2). If the scope change qualifies as a "not-in-kind" change, then it must go through the formal MOC process.

Once the decision has been made to include a scope change, the change activities are added to the *final work list* and the *live model* with clear identification and cross reference to the approved request.

At the end of the shutdown, all change activities can be extracted from the model and summarized separately for inclusion in the shutdown report. Should there be any disputes, a clear tracking process with the required approvals would be available for reference.

A sample flowchart is shown in Fig. 7.1.

7.2 Scope
7.2.0 Introduction and relevant PM processes

Phase 5 is the most critical part of scope management and includes a clear, structured process for any changes in scope.

The relevant standard processes are:

PMBOK	ISO 21500
5.5b Validate scope (new/additional work)	—
5.6 Control scope	4.3.14 Control scope

See Appendix A2 for details.

7.2.1 Additional and new work validation

The change control procedure must be strictly adhered to for additional and new work. This is outlined in Section 7.1.4.

7.2.2 Control of scope

Control of scope during execution is imperative, and the difference between scope growth and additional work must be clear. The **two** alternatives are clarified as follows:

The first is scope growth for identified work list activities.

> *Example "Replace four trays" becomes "Replace eight trays" after opening a distillation column. This represents additional work but is not outside the scope of the work list. Time and cost contingencies should cover this. Scope growth needs to be noted on the planning model for schedule and cost projections.*

The second is additional work: This category should be reserved for obvious oversights from the work list and necessary jobs uncovered during execution. They all need to go through the scope

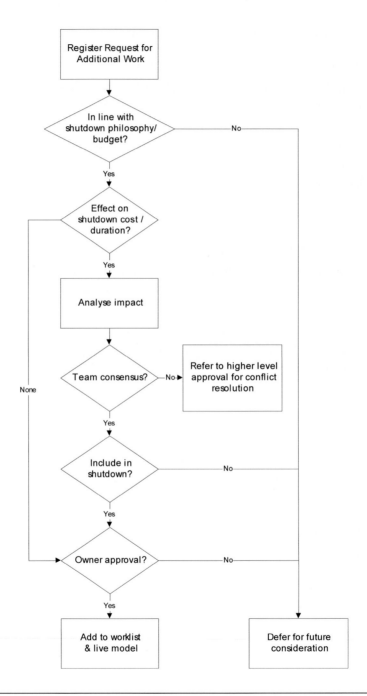

FIGURE 7.1

Scope change control process.

change control procedure and be **added to the planning model** (and separately identified) to **assess the impact on the critical path and resources available.** These items are outside the scope and therefore outside the shutdown budget—thus the need for **high-level approval**.

Quality control plans (QCPs) are generally the main source of change, generated when inspectors go inside vessels and furnaces and do close-up physical inspection. These QCPs all have to go through the process as described.

7.2.3 Scope creep

If scope control is poorly managed, there is potential for scope creep. This occurs when there is inadequate documentation for productivity, scope growth, and additional work. A well-structured critical path model will indicate where the problem lies. Three reasons can be given:

1. Productivity is low, i.e., planned work is taking longer than expected and/or
2. Actual work for an activity is greater than planned and/or
3. There is additional work that has not been added to the model.

The earned value graph has to be analyzed to determine what is influencing scope creep. This is discussed further in Appendix A4.4.

7.3 Schedule
7.3.0 Introduction and relevant PM processes

In phase 5 the turnaround work is carried out as per the schedule, which is determined by the baselined (frozen) critical path model. Actual schedule performance is measured against this model.

The relevant standard processes are:

PMBOK	ISO 21500
6.6 Control schedule	4.3.24 Control schedule

See Appendix A2 for details.

7.3.1 Progress monitoring

The critical path management (CPM) model would have been frozen at the beginning of phase 4, and the baseline, against which progress is measured, was set.

All scope growth items and additional work that have been approved by the scope review committee are resourced to ensure that the baseline critical path is maintained. They are then included in the live critical path model.

The planning engineer collects progress updates from the area supervisors and enters these into the critical path model on a daily basis. Critical path analysis determines if the project is ahead or behind the critical path schedule, and relevant reports are generated. This is completed prior to the daily progress meeting where all items requiring intervention are discussed, especially items that **have negative slack** (lag), i.e., they are extending the critical path duration. Options for **corrective action are discussed and initiated immediately after the meeting.** Box 7.1 gives an example of a change in critical path and the corrective action taken to get the project back on track.

Box 7.1: Bahrain refinery—critical path change during execution

On many of the older process units in Bahrain Refinery, complete heat exchangers are extracted and moved to the workshop where they are refurbished and hydro-tested before being reinstalled in the plant.

A crude oil process unit turnaround was planned where, in addition to refurbishing the heat exchangers, the stainless steel liner in a column had to be replaced. The critical path model identified a resource-restrained critical path for completion of the turnaround, as there were only a limited number of welders able to work in the confined space of the column. This resulted in the duration of the turnaround being based on the time required to complete the welding operation, while the heat exchanger process could be accomplished within this time allocation.

However, halfway though the turnaround, the critical path suddenly changed from stainless steel welding activities to replacement of heat exchangers. It was found that, although work had been completed on a number of exchangers, they were being stockpiled in the workshop due to transportation limits. Transport was immediately arranged for returning and reinstalling the exchangers into the unit, and within days the critical path returned to the welding-restrained activities.

Timeous intervention ensured that the turnaround was completed on schedule.

All approved scope growth items and additional work are discussed in the daily progress meetings and resourced to ensure that the baseline critical path is maintained. **Approved items are actioned promptly.**

This is illustrated in Box 7.2.

7.3.2 Monitoring and control activities

The area supervisor is at the forefront of controlling activities and ensuring that the plan is adhered to. The planning engineer interfaces with the area supervisor to ensure accurate feedback on progress to update the critical path model.

7.3.3 Saving a copy of the critical path model

Each day, after updates and critical path analysis, a copy of the critical path model is to be saved. At the end of the turnaround there will be a "slice in time" for each day of the turnaround that can be used as background to analyze the performance of the turnaround.

7.4 Cost

7.4.0 Introduction and relevant PM processes

Strict cost control is carried out on a daily basis in phase 5.

Box 7.2: Scope change after scope freeze—example (SAP/Primavera)

IAT: Inspection Advice Ticket.
IW32, IW41, KO88: SAP screens.
FMI: Inspection Manager.
PCR: Plant Change Request.
WO: Work Order.

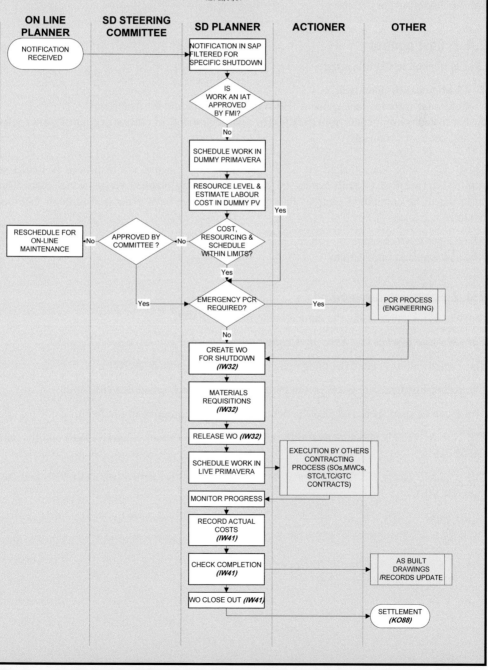

WORK FLOW
EMERGENT SHUTDOWN WORK
REV 25/04/04

The relevant standard processes are:

PMBOK	ISO 21500
7.4 Control costs	4.3.27 Control costs

See Appendix A4 for details.

7.4.1 Cost control

Cost is tracked in the following:

1. CPM model for man-hours
2. Enterprise resource management (ERM) materials module for materials
3. Estimated commitments from the ERM accounting module or other cost control software for staff and contract overheads.

These need to be consolidated on a daily basis, possibly using a spreadsheet or linked software application (see Box C18). S-curves are produced showing planned versus actual costs. The actual costs would be committed costs as they will not yet have been processed through the accounting module.

All changes to work scope are entered into the CPM model and identified as extra work with relevant cost and schedule implications.

7.4.2 Earned value method

Progress is measured using the earned value method (EVM).

Earned value is what you have physically gotten for what you have actually spent.

The earned value is called the budgeted cost of work progressed (BCWP).

The planned expenditure is called the budgeted cost of work scheduled (BCWS).

The actual amount spent is called the actual cost of work progressed (ACWP).

The cost variance is the numerical difference between the earned value (BCWP) and the actual cost (ACWP).

The schedule variance is the numerical difference between the earned value (BCWP) and the budget plan (BCWS).

Example

Plan

- *Clean 20 exchangers @ 100 h per exchanger*
- *Total planned hours: 2000*

Actual at time now

- *Planned to clean 12 exchangers at this time − 1200 h (planned value: PV)*
- *Only cleaned 10 exchangers − 1000 h (earned value: EV)*
- *Actual hours spent to date: 1250 (actual value: AV)*

Schedule variance (SV): EV − PV = 200 h behind schedule.

Cost variance (CV): EV − AC = 250 h overspent.

When setting up the project model, every cost must have an activity. Once this has been established, EV can be determined for each activity. The planning software can then take each activity in the same manner as the above example, to calculate **CV** and **SV** for all work to date, and, using a built-in cost algorithm and the CPM, predict the **final cost** and **completion date**. An EV graph can be generated, clearly showing these **four key parameters** and this can be used as a **compelling tool when reporting to management**.

Historical data indicates that once a project falls behind schedule and runs over budget at an early stage, **it is very difficult to recover**, unless immediate corrective action is taken and a concerted effort is made by the whole project team. One of the reasons is that it tends to cost more to speed up the schedule. For instance, overtime (money) is often required. In a turnaround environment progress needs to be monitored every day and immediate corrective action taken when necessary.

EVM is discussed in detail in Appendix A4.

7.5 Quality

7.5.0 Introduction and relevant PM processes

In phase 5, inspectors perform quality assurance and control. Primary focus is on containment of the pressure envelope. Welds and joints are therefore the most critical items requiring inspection.

The relevant project standard processes are:

PMBOK	ISO 21500
8.2b Manage quality	4.3.33 Perform quality assurance
8.3 Control quality	4.3.34 Perform quality control

See Appendix A8 for details.

7.5.1 Quality control plan recommendations

QCPs are generally the main source of change, as inspectors go inside vessels and furnaces and do close-up physical inspection. These QCPs all have to go through the process as described in Sections 7.1.4 and 7.2.2.

7.5.2 Tracking quality

Quality inspection entails the scrutiny, checking, and monitoring of critical elements of the work. Some examples are:

- Ensure that all QCPs are adhered to
- Check that all calibration certificates for testing equipment are genuine and valid
- Ensure that the breaking and making of joints are performed by competent persons using the correct tools and following the required method statements
- Ensure alignment of rotating equipment is undertaken by competent persons using the correct tools and following the required method statements
- Ensure that the internals of vessels, etc. meet process and inspection requirements and are free of debris before boxing up
- Ensure all welders are qualified for the types of welding that they are carrying out in accordance with records maintained by the inspection department.

The quality of welding is a significant factor. Tracking graphs for X-rays for each qualified welder will quickly identify those welders who are not performing to the required standards. Common practice in refineries is to X-ray 10% of carbon steel welds and 100% of alloy steel welds.

7.5.3 Hold points

Work on certain items of equipment requires approval by inspectors and/or process engineers before work can commence on the next activity. These "hold points" need to be included in the critical path model to prevent work continuing before previous work has passed inspection. Should this process not be followed, the quality control system will be compromised and provoke loss of control, or an omission might be discovered later resulting in rework. Items such as approval of weld X-rays are critical.

> *Example 1*
>
> *A welder may continue to do substandard work if earlier poor quality welds have not been X-rayed. If these inferior welds are not detected, the pressure system is put at risk, and should the unacceptable welds be discovered at a later stage, they will have to be cut out and redone.*
>
> *Example 2*
>
> *If internal inspection of a vessel is required and not done, and this is only identified after boxing up, then the vessel has to be reopened, which plainly results in wasted resources and time.*

Primary "hold points" include radiography, pressure testing, and internal inspection before a pressure vessel, furnace, or heat exchanger can be boxed up.

7.5.4 Pressure testing

Hydro, gas, and service tests are the most common. Any pressure envelope that permits water to be used for testing purposes should be tested at a prescribed elevated pressure for an assigned period using an approved pressure test gauge (in accordance with the appropriate applicable standard). Certain services require the water to be demineralized or chlorine-free.

Example

Chlorine attack on stainless steel systems is a real threat, and thus chlorine-free demineralized water is essential.

Air or nitrogen testing may be required for services where water is not permitted.

Example

Stainless steel pressure envelopes used for polyester polymer production require nitrogen testing.

Service testing is used for systems such as process control air systems.

Test certificates should be signed off by the appropriate authority immediately once the test has been approved.

7.5.5 Joints

The breaking and making of joints is vital with respect to containment. Critical joints that could delay start-up need particular attention and a joint check sheet is crucial to ensure that there are no leaks during start-up. Factors that may cause leaking include:

- Radial and axial alignment of the two flanges
- Incorrect gasket for a particular pressure rating or service
- Incorrect cold gap between flanges before tightening
- Incorrect joint tightening sequence
- Hot tightening not done to procedure (*example: steam turbine casing*)
- Poor flange face quality
- Competence of the fitter
- Knock-on effect of the tightening of one joint on other joints in the same piping configuration
- Use of old stretched bolts/studs.

Correct rating of spades (blanks/blinds) used in a turnaround is important. The process for the installation and removal of spades also needs to be well structured to avoid an incident. Box 7.3 describes an incident that initiated a process for spading and despading.

Jointing materials may not comply with the latest environmental requirements.

Box 7.3: Spading process example

On one occasion in the Bahrain Refinery, a spade was accidently left in a vapor line on a column after a crude unit turnaround. The column had to be rescaffolded to remove the spade, which delayed the start-up by a day.

Subsequently, the company initiated a process where all the required spades were hung against their respective tag numbers on a board where the plant was tagged for where spades were to be placed and the spades were progressively exchanged for the tags in the plant. During the turnaround the spade boards only had tags, and by the end of the turnaround, just prior to start-up, the spade boards only had spades.

Example

All low-pressure joint gaskets in Cape Town Refinery were originally made of compressed asbestos fiber (CAF). However, as a result of health concerns, all purchases of items containing asbestos were stopped in the 1980s. Consequently, the CAF gaskets were progressively replaced by spiral-wound metal gaskets.

Suitable tools include the "Equalizer" range of flange spreading, alignment, and pulling tools.[2]

7.6 Resources
7.6.0 Introduction and relevant PM processes

The application of resources to turnaround activities commences at the start of phase 5. Daily monitoring of resource usage and productivity takes place.

Relevant standard processes are:

PMBOK	ISO 21500
9.5 Manage team	4.3.20 Manage project team
9.6 Control resources	4.3.19 Control resources

See Appendix A9 for details.

7.6.1 Daily resource allocation

Resource buildup has to be in line with the planned resource histograms to achieve the required amount of work each day. Checks of planned versus actual resources for each contractor are required on a daily basis. Box 7.4 identifies a low allocation of resources up to the data date (time now), giving rise to an overloading of resources to complete the work in time.

7.6.2 Productivity

Following on from Section 5.6.5, actual productivity needs to be measured daily throughout the turnaround.

The critical path model will compare actual man-hours on the job to those that were planned. The time clocking system will record arrival on-site and departure from site. These need to be reconciled to determine actual productivity for the turnaround.

Once an individual clocks in he/she needs to change, obtain work permits, as well as tools and materials before proceeding to the work site. Tea and meal breaks must be allowed for, and, at the end of the shift, packing up and changing before clocking out needs time.

Firstly, the total actual man-hours spent on-site compared to that which was agreed with each contractor needs to be determined. Often the ramping up of resources at the start of a turnaround is slower than planned. Box 7.4 gives an example of lower than planned actual man-hours on the job.

Box 7.4: Sample resource histogram report

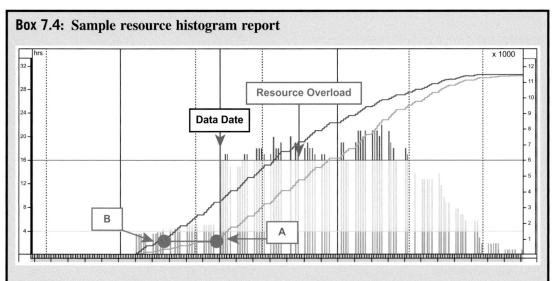

Comment

The vertical blue line depicts the "data date" sometimes referred to as "time now." Information to the left of the line is actual recorded workload (history) and information to the right of the line is the projected required workload. The cumulative curve to the left indicates actual man-hours completed (blue line) in relation to that planned (brown line). When the actual man-hours curve (blue line) is below the planned man-hours curve (brown line), the project is behind schedule. The horizontal line between the blue line at "time now" (point A) and the Brown line at point B indicates the amount of time behind schedule. The red part of the resource histogram to the right of "time now" is the required resource overload to get the project back on schedule.

The blue line curve could possibly be as a result of a slow actual ramp up of resources relative to that planned. Alternatively, the correct numbers are on-site but have not yet been applied to the work.

Secondly, the actual man-hours spent on-site contrasted with the actual man-hours on the job needs to be analyzed to determine changes to the planned factor of on-the-job time versus clocked time. A large discrepancy could be as a result of excessive time taken to obtain permits.

Thirdly, the ability to complete each activity according to the planned man-hours needs to be evaluated. Actual assessment at the "coalface" is needed to determine changes to the critical path. Earned value is commonly used here.

Example

The planned time for removal of a pump is 4 h. A fitter starts the job in his shift and spends 3 h on it. At the end of shift, he estimates another 3 h is required. The activity is therefore regarded as half complete. The estimated EV for partial completion of the removal of the pump is 3 h whereas the PV is 2 h. Schedule variance (EV−PV) is consequently 1 h behind schedule and labor cost variance is 1 h over budget.

Additional work, including rework and new work, needs to be logged and added to the critical path model using the scope change process, and not included into the above calculation.

The effect of scope growth on resources needs evaluation and the appropriate action must be taken to acquire additional resources (see Section 7.2.3).

Awareness of potential productivity inhibitors is essential. These include:

- Poor supervision and direction
- Work permit delays
- Waiting for construction equipment and tools
- Delays in material deliveries
- Inexperienced crafts
- Uncontrolled breaks
- Slow decision-making in the field
- Material misfits and rework
- Poor logistics planning and area conflicts
- Lack of information for proper field execution
- Adversarial contractual relations.

Productivity motivators include:

- Cordial supervision and craft relations
- Challenging work
- Proper work facilities
- Effective safety program
- Attractive incentive scheme and recognition for performance
- Efficient communications.

The application and availability of the most appropriate tools and equipment enhances productivity considerably.

Examples

- *Heat exchanger bundle pulling rigs*
- *Hydraulic bolt tightening*
- *Flange alignment tools.*

7.6.3 Fatigue monitoring

Working long hours causes fatigue, which could give rise to poor decision-making and actions that might endanger lives. It is therefore imperative that rest periods for certain activities be established.

Example

Air movement equipment is required to enable workers to work comfortably inside a hot pressure vessel, and frequent rest periods outside the vessel are needed.

7.7 Communication
7.7.0 Introduction and relevant PM processes

In phase 5, various daily communications take place. These include e-mails, reports, meetings, one-on-one discussions, radio communications, etc.

The **preparation for access,** both initially and at the start of each shift, using the permit-to-work (PTW) process, requires particular attention.

Related standard processes are:

PMBOK	ISO 21500
10.2b Manage communications	4.3.40 Manage communications
10.3b Monitor communications	4.3.39 Distribute information

See Appendix A10 for details.

7.7.1 Reporting

Reports should be clear, concise, comprehensive, and tailored to individual supervisors. The format of "48 h look ahead" report for an area helps each supervisor focus only on what has to be done in his area in the next few days. High-level reports should identify areas of concern.

Box 7.5 explains aspects of a typical graphical progress report.

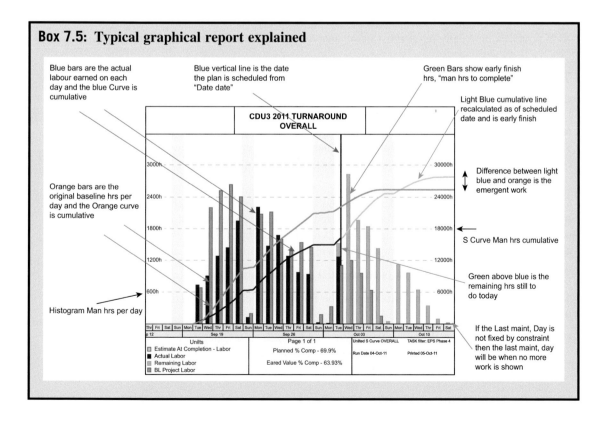

Sample reports are shown in Boxes 6.3, 6.4, and 6.5. Other report formats are discussed in Appendix C10.

Reporting and analyzing feedback from the field is crucial for correct decision-making with respect to corrective action. The way questions are asked about progress is important. For example, asking how complete an activity is often provokes an optimistic answer, whereas asking about the remaining duration generally prompts the field supervisor to think of all the things still to be done and elicits a more accurate answer. When the remaining duration is entered into the computer, it calculates the percentage that has been completed.

Analysis of progress occurs after the progress updates have been entered into the computer model and the critical path analysis has been performed. Various reports need to be reviewed to get an accurate picture of what is really happening.

Examples are:

- Activities being underresourced
- Poor productivity
- More work required on existing activities
- Additional work not being registered in the system
- Costs running over budget (based on commitments)

KPIs produced on a daily basis assist in the analysis (see Section 7.1.3).

7.7.2 Preparation for access

Once manufacturing has stopped, the process system is flushed and furnaces are cooled in preparation for access. Methods of flushing vary depending on the process fluids involved. Chemical decontamination could be used to speed up the process to get quicker entry, although there might be a higher risk associated with this method. The flushing of lines is often referred to as "sweetening." Oil refineries commonly use diesel, steam, and air whereas some areas require an inert gas such as nitrogen to prevent pyrophoric combustion.

Once flushing has been completed, vessels and furnaces are spaded off (physically isolated), electrically disconnected, opened, and force ventilated for cooling and oxygenating. Various tests are then carried out. Oxygen, flammability, and toxic gas testing is essential. Other tests such as benzene level checks, etc. are completed to ensure that they are within safe limits before an entry permit is issued.

The **permit-to-work** issue system is an essential part of the shutdown process.

The rules for permit issue vary, but issue of the following is common practice in the petrochemical industry:

- **Hot work permit** for the relevant area covered by each shift, extendable for up to 24 h and renewed daily on condition that there are sufficient fire watchers present at all times and that all potential emission sources are sealed. Hot work is generally defined as any work that could cause ignition.
- **Crane lift** permits for specific lifts.

- **Excavation** permits for particular excavations for the shift, extendable for up to 24 h and renewable daily on condition that regular gas tests are carried out for deeper excavations—the presence of a hole watcher may be required.
- **Confined space entry** permit for a particular confined space (vessel, boiler, etc.) After the required tests have been completed successfully, a hole watcher is appointed and the necessary breathing apparatus and rescue harnesses made available at the access point. Regular air testing is normally required throughout the shift with the permit extendable up to 24 h and renewable daily.
- **Special hazard permit** for radiographic inspection, asbestos removal, chemical handling, and leaded tank work.
- **Electrical work** permit for all electrical work.

If the permit rules are not adhered to, major delays can occur, and major incidents could result.

7.7.3 Daily supervisor responsibilities

Ideally, each supervisor would have been intimately involved during the planning phase of all the work in his area of supervision (see Section 5.6.2). He would spend a significant amount of time observing work and communicating with the people who are performing the work. Much of the communication is nonverbal, in that the supervisor often needs be very aware of, and sensitive to, worker attitudes, fatigue, and confidence in completing activities on schedule.

Documenting observations in the daily progress reports and the shift turnover book is important, both for accurate prediction and continuity of work. It is also essential to follow the written program derived from the critical path model as deviation could cause a serious impact in another area.

Examples

1. *A crane rigging supervisor, responsible for servicing the whole site, chose to complete the "nice-to-do" crane rigging jobs first, all of which had large amounts of slack on the critical path. When he got to the other activities with little slack, he found he had insufficient resources and this process actually pushed the crane rigging activities onto the critical path.*
2. *A workshop supervisor in charge of heat exchanger inspection and testing, stockpiled exchangers in the workshop after completion of testing. This resulted in the field installation teams running low on work during the turnaround and then getting a flood of exchanger installations toward the end, putting them on the critical path (see Box 7.1).*

Written guidance in the form of method statements, specifications, drawings, materials requisitions, purchase orders, QCPs, etc. are essential for the supervisor to maintain the required quality standards.

Daily critical path analysis, resourcing, and costing must ensure that the decision makers are fully informed on a strategic level, in order to be able to take any necessary corrective action quickly and decisively. The correct reports need to be distributed to the correct people to ensure this.

7.7.4 Daily activities of the turnaround project manager

The turnaround project manager must have "a finger on the pulse" hour by hour when on-site. When off-site, he should be available on "24 h" callout.

A typical routine is outlined in Box 7.6.

Box 7.6: Typical daily activities of a turnaround project manager

The typical day-to-day to-do list of a turnaround project manager would look something akin to this:

1. Check previous 24 h progress with planning engineer
2. Check cost control and forecast with planning/cost control engineer
3. Visit safety portacabin to check safety matters
4. Visit stores to check stores/materials problems
5. Visit workshops to check workshop issues
6. Tour site to talk to people to:
   a) Verify progress as discussed with planning engineer
   b) Verify safety issues as discussed with safety officer
   c) Assess overnight work progress
7. Visit permit office to check permit-to-work issues
8. Check quality issues with inspectors
9. Meet area engineers before daily turnaround progress meeting
10. Chair daily turnaround progress meeting and allocate actions for resolution within the following 24 h
11. Resolve all issues that cannot be handled by members of his team
12. Report progress to decision executive.

7.7.5 Daily work progress meetings

Daily meetings are an excellent forum for reviewing progress and initiating corrective action. All interested parties need to attend these meetings—operations, maintenance, engineering, process engineering, inspection, purchasing, planning, etc. Minutes should be issued within hours of these meetings.

The agenda must focus on:

- Health, safety, and environment issues
- Critical path progress—ahead or behind
- Additional and new work
- Corrective actions.

7.7.6 Conflict resolution[4–6]

Conflicts inevitably arise during the turnaround execution phase as everyone is under great pressure to meet the target of completing the work within the agreed duration. The logical approach would be for all parties to agree to work toward a common goal, and then to focus on the problem and not the people involved. Steps to follow could be to define the nature of the conflict, define the options and consequences of each option, and agree on a solution. If agreement cannot be obtained, the issue has to be raised to the next level of management. However, team consensus is always the preferred option.

Between client and contractor
A well-worded contract is the starting point. Reasonableness and common sense are required to achieve the common goal of completion in the agreed time. A cordial previous long-term relationship between the client and contractor often builds trust and so minimizes this sort of conflict.

Between contractors

Typical conflicts arise from:

- One contractor delaying another
- Two contractors working in a close, confined space together
- One contractor undoing another contractor's work

The use of hold points for signing off on work done before the next stage of work is started, should be applied meticulously.

> *Example*
>
> *A welder completes welding during the day shift and the welds are to be X-rayed on the night shift. However, the required X-raying is not completed on the night shift and an insulator starts insulating the next day before the welds have been approved. This could cause further delays and require removal of the insulation for X-raying and possible rewelding.*

Between supervisors

Area supervisors using common resources may be at "loggerheads" about who gets the common resource first. The priority should be set in the critical path model and adhered to.

Elevation

If an issue cannot be resolved between two individuals, it has to be raised to the turnaround project manager before the issue escalates. If the turnaround project manager is not able to resolve the problem, then he needs to approach the decision maker.

7.7.7 Flawless turnaround awareness program

A flawless turnaround awareness program is initiated at the start of this phase. It entails health, safety, environment, and quality, and promotes the minimization of related incidents. The primary KPIs and targets are **incident-free turnaround** and **zero start-up incidents**.

Section 7.8.3 discusses risks related to handover and start-up.

7.7.8 Recording of interiors of vessels and key activities

The use of videos and still photography is very helpful to record any damage found when opening up equipment, as well as recording any repair work. These records are invaluable when planning future turnarounds, especially to those who were not involved with the turnaround when the damage was detected.

Images of the interior of furnaces, boilers, and fluid catalytic cracking (FCC) units, and photos and videos of major maintenance work, such as FCC cyclone replacement, are most helpful for planning any major work required in future turnarounds.

7.8 Risk (including HSE issues)

7.8.0 Introduction and relevant PM processes

In phase 5 the primary focus is on risk mitigation. Any emerging risks have to be addressed promptly and the risk register continuously referred to and updated.

Related standard processes are:

PMBOK	ISO 21500	ISO 31000 risk management
11.2 Identify risks	4.3.28 Identify risks	1. Establish the context
—	—	2. Communicate and consult

11.3 Perform qualitative risk analysis
11.4 Perform quantitative risk analysis
	4.3.29 Assess risks	3. Risk assessment
		  Risk identification
		  Risk analysis
		  Risk evaluation

11.5 Plan risk response 11.6 Implement risk responses	4.3.30 Treat risks	4. Risk treatment
11.7 Monitor risks	4.3.31 Control risks	5. Monitor and review

See Appendix A7 for details.

7.8.1 Health, safety, and environment oversight

Daily HSE walkabouts by the HSE staff are mandatory. Unhealthy, unsafe, or polluting acts are reported in the daily progress meeting. STOP card data and statistics are also reported in this meeting. Meticulous attention to detail is required for certain activities, especially for access to confined spaces: gas testing, hole watch, etc.

Section 6.7.3 describes the HSE relationships and Section 6.8.1 describes Safety, Health Environment and Quality (SHEQ) control.

7.8.2 Addressing known unknowns

A contingency allowance is used to address known unknowns, that is, items that have been known to occur but might not actually materialize.

Box 7.7 describes a known unknown that got out of hand.

Box 7.7: Welder strike

A turnaround at Chevron Cape Town Refinery in the early 2000s came to a standstill when the welders on-site went on strike for more pay.

Comment

The relationship between the contractors and the union had not been cordial even before the turnaround, and this should have been addressed as a key component of stakeholder relationships. Alternatively, nonunion workers could have been employed for the turnaround.

7.8.3 Handover and start-up

Flawless start-up is a KPI that is valid until the plant is producing on-spec product at the required rate. As noted in Section 3.8.3, **the risk profile for the plant is at its highest during start-up**. In the oil and gas industry, handover and start-up risks primarily relate to hydrocarbon release.

An example of a disaster during start-up is recounted in Box 7.8.

Box 7.8: CSB report—BP Texas City Refinery start-up disaster[3]

Accident: BP America Refinery Explosion.

Location: Texas City, TX.

Accident Type: Oil and Refining—Fire and Explosion.

At approximately 1:20 p.m. on March 23, 2005, a series of explosions occurred at the BP Texas City refinery during the restarting of a hydrocarbon isomerization unit. Fifteen workers were killed and 180 others were injured.

Joint integrity

Joint integrity is of prime importance to ensure the pressure envelope is not compromised. Final removal of blinds after internal inspection or hydro-testing and the sealing of the final joint is discussed in Section 7.5.5.

Pre start-up safety review

A **pre start-up safety review** (PSSR) is mandatory to ensure that the return to operations will be safe and seamless. This includes a walk through the plant tracing all lines. Some of the key items to note are:

1. All blinds removed (see Box 7.3)
2. All instruments reconnected
3. All control valves stroked
4. All relief valves reinstalled with documented testing completed
5. All modified process lines and equipment are as per approved piping and instrument diagrams (P&IDs)
6. All pressure testing complete and documented
7. All drawings marked up to reflect the "as-built" status
8. Operator training for modifications in progress
9. Maintenance training for modifications in progress
10. Start-up spares available
11. Physical check inside all vessels before boxing up
12. Correct **thermal insulation** reinstated where required.

   Example

   A petrochemical plant was started up with insufficient thermal insulation on the thermal heating oil circuit. The product would not polymerize and the plant had to be shut down and more thermal insulation applied before being started up again. The delay in achieving production of quality product was at least a week.

The MOC process also requires that all documentation, including drawings, etc., be updated to the "as-built" situation.

Technical (design and process engineering) and maintenance assistance and training support are required during the start-up period and until the process unit is online and stable.

The PSSR is discussed in detail in Appendix D8.

An example of a PSSR checklist for the end of a normal refinery turnaround is shown in Box 7.9.

7.8.4 Commissioning and start-up of new plant and equipment

New plant and equipment may have been installed during the turnaround. The start-up procedure for any new items would have been drafted before initial start-up and further amended based on subsequent experiences of this initial start-up. Items listed in the PSSR in Section 7.8.3 should to be covered. Other examples of precommissioning activities are:

- Motor direction check
- Instrument loop check.

A punch or "but" list is then created. This is a list of all items to be repaired before moving into the commissioning stage. This could include details such as incomplete thermal insulation, missing handrails, etc.

Prior to the commissioning stage, a mock operation is simulated, to mimic normal operations as closely as possible. This includes:

1. Process: pump around of product
2. Services: start-up of services and circulation of steam, water, air, nitrogen, oxygen, etc.

Commissioning and start-up of new plant and equipment is discussed in detail in Appendix D.

7.9 Procurement
7.9.0 Introduction and relevant PM processes

In phase 5, intensive management of all turnaround execution contracts takes place. All changes have to go through the change control process described in Section 7.1.2.

Related standard processes are:

PMBOK	ISO 21500
12.3 Control procurements	4.3.37 Administer procurements

Box 7.9: PSSR checklist for refinery turnarounds

Process startup safety review

Process unit:	
Description of modifications made during turnaround	

Description of tasks	*Initials/date*
1. Remove all unnecessary maintenance materials from plant including the following: 　a) Unneeded staging/scaffolding 　b) Debris: tags, insulation, gaskets and other debris 　c) Hydro-test equipment, temporary hoses, and all other temporary related equipment 　d) Temporary piping connections 　e) Replaced test gauges.	
2. Perform a fire and safety equipment inspection using a checklist. Submit a signed-off copy to support supervisor and attach a copy of this report. 　a) Replace missing or inoperable equipment.	
3. P&ID check any equipment worked on during this turnaround. While doing P&ID check: 　a) Install all missing bull plugs 　b) Make sure all valve position are correct 　c) Review blind list.	
4. On all flanges that were worked on: 　a) Hammer test bolts 　b) Check for proper gaskets 　c) Sign blind list and forward to support supervisor.	
5. Perform alarm checks per pre-start-up alarm checklist.	
6. Conduct pressure test for tightness as per procedures.	
7. Loop check all instruments and verify that: 　a) All control valves stroke properly and fail in proper position (fail close or fail open), and 　b) All valves are open to instrument bridles.	
8. Check that all temperature indicators are working (ambient test)	
9. Review valve repair list. Check each valve worked on for the following: 　a) Packing adjustment 　b) Adequate amount of packing 　c) Valves in correct position.	
10. Review pressure relief device list and check that each valve is in service: 　a) ID Tags are in place 　b) Block valves on Pressure Relief Device (PRD) blocks are open 　c) Chains are installed and locked on PRD blocks 　d) Blocks have been positively verified open by visual inspection or X-ray 　e) Valve is not overdue for testing that requires shutdown.	
11. Review the gauge glass repair list and check that the gauge glass valves are in correct position.	
12. Review turnaround records and verify that all equipment "head up" sheets have been signed and archived.	
13. Make certain the operating and maintenance procedures have been modified and approved through the management of change process to reflect changes made in the process. 　a) Verify that current start-up procedures are available to operators.	
14. Review work list with turnaround team, maintenance, inspection, and engineering to assure completion of critical jobs.	

Maintenance review/date	Support supervisor review/date
Engineering review/date	Operations review/date
Inspection review/date	Other review/date

Description of work performed or equipment installed

See Appendix A9 for details.

7.9.1 Contract management

Contract management is a natural extension of the change process in Section 7.1: *Integration*. The change process, agreed at the negotiation stage, needs to be strictly enforced.

The client's representatives need to:

1) Closely monitor the contractors' performance against agreed targets and ensure that immediate steps are taken to rectify any issues, with emphasis on quality and time.
2) Ensure that the client organization does not delay the contractor, especially with respect to timely issue of permits to work.
3) Minimize the knock-on effect of one contractor delaying another by effectively coordinating activities.
4) Ensure all variations and emergent work requests are approved and issued in written form in a timely manner.
5) Emphasize to client staff and contractors that work will not be paid for if no written instruction is issued to the contractor.
6) Validate contractors' claims timeously.
7) Sign off completed work as soon as possible.

7.10 Stakeholder management
7.10.0 Introduction and relevant PM processes

In phase 5, it is of paramount importance to keep the sponsor/decision executive informed of progress (ahead or behind schedule) on a regular basis. If progress is being impeded, the sponsor/decision executive can initiate contingency plans if necessary (see Box 5.15).

For gas plant start-ups, excess flaring occurs and the relevant environmental authority and surrounding industries may need to be informed.

Related standard processes are:

PMBOK	ISO 21500
13.3 Manage stakeholder engagement 13.4 Monitor stakeholder engagement	4.3.10 Manage stakeholders

See Appendix A10 for details.

7.10.1 Progress reporting

Regular progress reporting to the decision executive/owner will prevent unpleasant surprises.

7.10.2 Start-up coordination

Detailed coordination with key stakeholders, especially the decision executive/owner and the operators of the plant, needs to commence well before actual start-up of the plant. This may entail frequent meetings just before start-up.

A PSSR needs to be signed off by the decision executive/owner before start-up commences.

Notifications to interested government and regulatory authorities are listed in Section 6.10.2.

7.11 Summary: phase 6—shutdown execution
7.11.1 Focus areas

1. Process unit shutdown and inspection and repairs
2. Daily meetings
3. Daily schedule reviews and update
4. Daily cost tracking and reporting
5. MOC requirements documented
6. Additional work review and processing
7. Repair documentation (QCPs)
8. Pre-start-up safety checks and a PSSR
9. Start-up process
10. Cleanup unit

7.11.2 Deliverables

- Shutdown execution completed as per plan and within budget
- Incident-free start-up

7.11.3 Tools and processes

1. Planning model
2. Shutdown guidelines/procedures
3. Additional work approval process

References

1. Kreiling J. Seven refineries, one solution. *Primavera Magazine*. Available from: www.primavera.com.
2. Equalizer International Group. Available from: www.equalizerinternational.com.
3. US Chemical Safety and Hazard Investigation Board (CSB) report. Available from: www.csb.gov/bp-america-refinery-explosion/.
4. Fisher R, Ury W. *Getting to yes*. Random House; 1991.
5. Ury W. *Getting past no*. Random House; 1991.
6. Stone D, Patton B, Heen S. *Difficult conversations*. Penguin; 2000.

Turnaround project closure

Phase 6 Post turnaround

8.0 Outline

This phase starts once the turnaround has been completed and the plant is handed back to the operator. It is considered the highest risk period during the life of the plant (see Section 3.8.3). At the same time, cleanup and demobilization activities are completed. Final accounting, preserving of the actual critical path method model for the turnaround and lessons learned are also cataloged during this phase.

Activities are discussed under the following headings:

1. integration
2. scope
3. schedule
4. cost
5. quality
6. resources
7. communication
8. risk (including health, safety, and environment [HSE] issues)
9. procurement
10. stakeholder management

Specifically Table 8.1 references PMBOK and ISO 21500 processes applicable in this phase.

Table 8.1 Phase 6—post-turnaround.

No	PMBOK	ISO 21500
1	4.6 Close project	4.3.7 Close project
	4.2 Manage project knowledge	4.3.8 Collect lessons learned
7	—	4.3.39 Distribute information
10	13.3 Manage stakeholder engagement	4.3.10 Manage stakeholders
	13.4 Monitor stakeholder engagement	—

Turnaround Management for the Oil, Gas, and Process Industries. https://doi.org/10.1016/B978-0-12-817454-8.00008-3

8.1 Integration

8.1.0 Introduction and relevant project management (PM) processes

All too often companies fail to spend the necessary time and resources to collect and evaluate valuable information from recently completed turnarounds in preparation for the next turnaround. Pacesetters ensure that this last and very important phase comes to a timely and effective close. A biding mantra is: "We'll do better next time."

The relevant standard processes are:

PMBOK	ISO 21500
4.6 Close project	4.3.7 Close project
4.2 Manage project knowledge	4.3.8 Collect lessons learned

See **Appendix A1** for details.

8.1.1 Overview

Post-turnaround is the final (sixth) phase of the turnaround development process. This phase includes unit cleanup, demobilization, performance evaluation, documentation, closing out contracts, final cost reports, lessons learned, and the final turnaround report and closing out of the project. A post-turnaround audit is required and included in the final turnaround report. The phase is complete when the final turnaround report has been presented to the decision executive and this report then becomes the starting document for the next turnaround.

The critical path method (CPM) model showing actuals versus planned is required to be archived.

Post-turnaround activities should be completed within 1–2 months after the turnaround.

8.1.2 Detailed analysis of key performance indicators

The following should be analyzed:

1. Schedule
   a. Behind or ahead of critical path: hours
   b. **Schedule variance** (SV): actual duration versus planned duration
   c. Schedule performance index: an alternative to SV
      i. Deviation from planned: earned value/planned value
2. Cost
   a. **Cost variance** (CV): actual final cost versus planned final costs
   b. Cost performance index: an alternative to CV
      i. Deviation from budget: earned value/actual cost
   c. Deviation from budget: estimate at completion/budget at completion
   d. Additional work: actual versus contingency %

3. Scope

    a. Emergent work

        i. Man-hours for emergent work as % of total turnaround man-hours: % (minimize)

4. HSE

    a. Lost time incidents: number (target: zero)

    b. Flawless start-up (no incidents, **no leaks**): number (target: zero)

5. Productivity

    a. Productivity Index

        i. Actual total man-hours on the job/total clocked man-hours.

Vendor and contractor costs would be committed costs rather than what was actually paid to the vendors/contractors (see Appendix G: Commitment).

Should the planned versus actual critical path model have been saved on a daily basis, then day-by-day occurrences can be checked and analyzed. Causes of variances are important to prevent these issues arising in the next turnaround.

8.1.3 Overall assessment of performance

Once the relevant information is available, a formal review of the turnaround is required. Mistakes as well as successes need to be examined to ascertain what went wrong (or right) and why and how it happened. Assessment focus could be carried out in three categories: organization and systems, staff performance (see Section 8.6.1) and contractor/vendor performance (see Section 8.9.1). Organization and systems performance discussions should include:

- Effectiveness of the turnaround management organization
- Quality and caliber of staff required for turnarounds
- Level of turnaround staffing and timing of assignment to the turnaround, i.e., whether the staff were assigned to the turnaround at the right time or too late or too early
- Organizational interfaces between the turnaround project team and the rest of the organization, as well as other stakeholders
- Audit findings.

8.2 Scope
8.2.0 Introduction

In phase 6 the work that has been carried out has to be analyzed: actual versus planned and additional work versus planned.

8.2.1 Statistics

A full description of the actual scope of work carried out is required and must include an analysis of planned versus actual work done without any additional work as well as details of all additional work. This should be obtained from the CPM model. Categorization and analysis of the additional work is critical for determining lessons learned so that additional work can be reduced and/or anticipated for the next turnaround.

8.2.2 Reallocation of outstanding work

Outstanding work comprises:

1) Planned work that was not completed
2) Additional work that was not completed
3) Work identified for onstream maintenance
4) Work identified for the next turnaround.

Planned and additional work would generally be scheduled for an opportunity shutdown or the next turnaround. Work identified for onstream is added to the onstream work list. Work identified for the next turnaround would be logged as such in the Computerized Maintenance Management System (CMMS).

8.3 Schedule
8.3.0 Introduction

Primary turnaround closure activities include:

I. Completion of punch list items
II. Mechanical, instrument, and electrical support for start-up where immediate attention to leaks and repairs is essential.

In phase 6 detailed analysis takes place as to actual progress against schedule. All scope growth and additional work should be analyzed in detail to ascertain lessons learned for input into planning the next turnaround.

8.3.1 Statistics

Key performance indicators (KPIs) are analyzed as well as reasons for the performance (either good or bad).

8.3.2 Schedule reports

Further to Section 7.3.3, reporting of actual progress against schedule would have been done on a daily basis and the model saved each day. Daily schedule reports are to be collated, analyzed, and summarized.

8.4 Cost
8.4.0 Introduction

After the turnaround has ended all costs must be collected, collated, and analyzed. Costs due to scope growth as well as additional work need close scrutiny and analysis to produce lessons learned for use in planning the next turnaround.

8.4.1 Cost reports

Once all materials have been disposed of, and contractors claims have been addressed, it is time to freeze the turnaround account.

If the earned value method has been used, the generation of cost reports should be very simple. All extra work carried out could be extracted from the model with categorized reasons for these extras.

The use of complementary software for contract control and the dumping of data to spreadsheets simplifies the analysis and reporting process (see **Appendix C12**). Graphs are important tools for presenting the final cost report. Commitment costs must be reconciled with actual final costs.

It may take time to complete final payments for major work carried out in the turnaround (see Section 8.9.2).

A supplementary expenditure request might be required for the extra work carried out and a structured management of change (MOC) process would make the compilation of this expenditure request straight forward.

8.5 Quality
8.5.0 Introduction

In phase 6 all quality records from the turnaround are analyzed. Data is extracted for reporting and lessons learned records.

8.5.1 Reports

All quality reports need to be completed and archived. All maintenance and inspection databases require updating.

Evidence such as X-rays and post-weld heat treatment graphs must be filed with their respective reports.

8.5.2 The value of quality records

Quality records are critical for assessment of what needs to be achieved in the next turnaround.

Examples

1. *Thickness gauging measurements over time determine the rate of wastage (corrosion/erosion) and can predict when a new pressure vessel needs to be ordered, possibly for the next turnaround.*
2. *The state of the internals of a vessel determines the amount of work required when opening that vessel again.*

8.6 Resources
8.6.0 Introduction

Phase 6 includes performance reviews and demobilization. In addition, the final cleanup is completed and excess material and redundant equipment removed from site.

8.6.1 Performance review

Basic criteria for performing team members' performance includes:

  i. Leadership and team player qualities
 ii. Quality of work
iii. Attitude and work ethic
 iv. Technical ability
  v. Cost consciousness
 vi. Communication skills
vii. Contribution to the success of the turnaround.

As part of a self-appraisal, these are some of the questions that could be asked:

a) What did I do well?
b) What could I have done better?
c) Was I a good leader?
d) Was I a good team player?
e) Did I add value to the turnaround planning and execution?
f) What could I learn from the experience?

It is common practice for staff to receive bonuses if the turnaround has exceeded targets, especially for duration, cost, and zero incidents.

8.6.2 Unit cleanup

Unit cleanup includes:

1. Removal of all scaffolding
2. Removal of all temporary installations/connections
3. Replacement of handrailings, floor gratings, safety barriers, etc.
4. Removal of mobile equipment
5. Removal of all materials and waste
6. Repainting
7. Final thermal insulation cladding.

These activities would be carried out while the plant is starting up, adding to what is already a **high-risk period.**

8.6.3 Demobilization

Teams required for start-up are identified before the contract staff is demobilized. Return of security clearances needs close attention to ensure no unauthorized access is gained in the future.

Turnarounds are very stressful, often frantic, events, and it stands to reason that core team members will be drained and exhausted in the aftermath. Many will be looking forward to taking time off before being assigned to their respective departments. However, there will inevitably be a lot of loose ends that will need to be tied up. It would be astute for a few key team members to finish the paperwork before going on leave.

8.6.4 Return of high-value hired equipment

Hire costs of dormant high-value equipment, such as mobile cranes, fixed tower cranes, high-pressure cleaning equipment, welding machines, mobile compressors, post-weld heat treatment rigs, etc., can escalate quickly and should be removed the moment they are no longer required on-site. Evidence of the time and date of removal must be retained for payment purposes.

8.6.5 Disposal of excess materials and equipment

Surplus materials are returned to stores and credit obtained.

Auctioning of redundant equipment is carried out.

The disposal of excess equipment and materials must be resolved before the cost reports are finalized. There could be several reasons for having excess material.

Example

Typically, craftsmen withdraw extra consumables from stores if the stores are far from the work-site. To reduce this problem, a satellite store for consumables and tools could be established close to the work-site.

Equipment that has been dismantled and will no longer be used could be sold to similar plants, although some might only have a scrap value. The plant asset register and other records have to be adjusted accordingly. Unused materials could be sold back to the vendor or placed back into stock.

8.7 Communication
8.7.0 Introduction and relevant PM process

In phase 6, clear communication is required both for start-up and to optimize the demobilization and removal of equipment once these are no longer required.

Documentation of lessons learned and imparting the awareness of these lessons is vital. The turnaround report winds up the project and must include the findings of the final audit report and lessons learned.

Related standard process is:

PMBOK	ISO 21500
–	4.3.39 Distribute information

8.7.1 Demobilization process

Demobilization entails packing up and removing the contractor from site. All equipment, redundant materials, and waste are cleared. Good communication ensures acceptance of a clean, clear site by the client/operator.

8.7.2 Documentation

> The job is not finished until the paperwork is done.

Documentation includes:

1. Producing as-built drawings
2. Updating process operating procedures
3. Updating maintenance procedures
4. Updating equipment historical records and asset register
5. Updating inspection records including the filing of process safety information such as relief valve and hydrostatic test certificates
6. Correcting materials templates in anticipation of the next turnaround
7. Completion of MOC process
8. Saving the turnaround planning model and planning templates for use in the next turnaround
9. Listing recommendations for onstream maintenance.

8.7.3 Lessons learned

Lessons learned should be summarized from the following:

1. Planning model
2. Shift turnover book
3. Cost reports
4. Contractor evaluations
5. Other records.

If a problem was encountered during the turnaround and not adequately resolved at the time, an investigation should be carried out so that a better solution can be recommended for the next turnaround. Brainstorming and dedicated sessions for root cause failure analysis or fault tree analysis could be very helpful.

8.7.4 Application of lessons learned

Guidelines for application of lessons learned:

a) The turnaround report should be compiled in a way that makes it easy to extract lessons learned for application in the next turnaround
b) The Pareto 80:20 rule should be applied to focus on primary issues for improvement in carrying out the next turnaround

c) The planned and actual activities from the completed turnaround must be saved for future reference and improvement.

A structured improvement process is discussed in Section 9.3.

8.7.5 Turnaround report

The turnaround project manager should ensure that the input into the report is candid, honest, and useful to ensure improvement in the performance of future turnarounds.

The **turnaround report** needs to be as comprehensive as possible with reference to other records where required. This document is the starting document for the next turnaround and it is important to learn from the experience in order to improve the next time. This document is also a reference for work to be done online between turnarounds. In compiling the turnaround report, the following should be included:

1. Introduction, including turnaround philosophy, turnaround dates, duration, critical path, budget, and major capital work done
2. Detailed reporting by area, equipment type, and/or discipline
3. Cost analysis, including extras and categorized reasons
4. Planning and resourcing
5. Communication
6. Logistics
7. Technical
8. Capital and major maintenance work
9. Safety, health, environment, and quality (SHEQ)
10. Contractor performance
11. Start-up
12. **Lessons learned** (summarized from 2 to 11 above)
13. List of support personnel and organizational structure
14. Recommendations for online operation
15. Recommendations for the next turnaround.

The above should include the KPIs so that they are easily extracted for benchmarking. They would possibly include:

1. Schedule variance (deviation from the target date)
2. Cost variance (deviation from the budgeted cost)
3. Incident rate (SHEQ lost time/recordable incidents)
4. Scope freeze date (number of days before the turnaround start)
5. Work order (scope) growth from expenditure request submission date
6. Run time since last planned turnaround
7. Number of unplanned shutdowns since the last planned turnaround and the duration of each
8. Total labor man-hours planned and actual with extra work separated
9. Labor man-hours for planned critical path activities

10. Peak planned and actual resourcing by trade
11. Start-up duration—planned and actual

An example of the contents of a turnaround report is shown in Box 8.1.

The **final turnaround report** is to be presented to the decision executive and widely distributed.

Box 8.1: Turnaround report template[1]

1. Objectives
   - **1.1.** Cost
     - **1.1.1.** Number of work orders
     - **1.1.2.** Direct labor man-hours
     - **1.1.3.** Total cost
   - **1.2.** Schedule
     - **1.2.1.** Number of activities
     - **1.2.2.** Elapsed days: oil out to oil in
     - **1.2.3.** Elapsed days: turnaround
     - **1.2.4.** Lost time (weather, etc.)
     - **1.2.5.** Average manpower staffing
   - **1.3.** Quality
     - **1.3.1.** Testing guidelines
     - **1.3.2.** Amount of rework
     - **1.3.3.** Routine repair items
       - **1.3.3.1.** Towers and vessels
       - **1.3.3.2.** Heat exchangers
       - **1.3.3.3.** Rotating equipment
       - **1.3.3.4.** Electrical/instrumentation
   - **1.4.** Inspection
     - **1.4.1.** Extent of inspection
     - **1.4.2.** Inspected items not repaired
   - **1.5.** Safety
     - **1.5.1.** Tagging
     - **1.5.2.** Workplace tidiness
     - **1.5.3.** Accidents
2. Execution
   - **2.1.** Organization
     - **2.1.1.** Philosophy
     - **2.1.2.** Company personnel assignments
     - **2.1.3.** Contractor personnel assignments
   - **2.2.** Turnaround management systems and controls
     - **2.2.1.** Turnaround management system
     - **2.2.2.** Progress communication/feedback
     - **2.2.3.** Daily progress tracking
     - **2.2.4.** Consultants assistance
   - **2.3.** Staffing
     - **2.3.1.** Schedule
     - **2.3.2.** Shift length, work week
     - **2.3.3.** Manpower peak

Box 8.1: Turnaround report template[1]—cont'd

2.4. Contractors
  2.4.1. General contractor
  2.4.2. Subcontractors
2.5. Inspection
  2.5.1. General
  2.5.2. Extra work generated from inspection reports
  2.5.3. Recommendations for future turnarounds
2.6. Tools and equipment availability
  2.6.1. Handling equipment
  2.6.2. Cranes
  2.6.3. Tool cribs
2.7. Materials supply
  2.7.1. General
  2.7.2. Gaskets, bolts, studs, etc.
2.8. Weather and other work interruptions
  2.8.1. Weather
  2.8.2. Upsets
2.9. Quality of workmanship
  2.9.1. Field welds
  2.9.2. Fugitive emissions
  2.9.3. Bolts, studs
3. Conclusions and recommendations
  3.1. Management systems and controls
  3.2. Management staffing
  3.3. Purchasing/warehousing
4. Exhibits
  4.1. Executive summary
  4.2. Organization chart
  4.3. Man-hour breakdown
  4.4. Turnaround summary schedule
  4.5. Completion dates
  4.6. Final turnaround detailed progress report
  4.7. Final turnaround summary progress report (curve)
  4.8. Final turnaround manpower
  4.9. Final earned versus burned report
  4.10. Performance analysis report
  4.11. Inspection reports
  4.12. Final turnaround safety report
  4.13. Unit turnaround history

Adapted from BP Texas City Refinery (turnaround report format).

8.7.6 Archiving documentation

The archiving of documentation for easy retrieval for the next turnaround is essential. Documents include:

I. Inspection records, including completed test sheets, X-ray records, etc.
II. Maintenance records, including equipment repair records, etc.

III. As-built drawings
IV. Turnaround network templates
V. Cost reports
VI. Contract files
VII. Material templates for each item of equipment.

8.8 Risk (including HSE issues)
8.8.0 Introduction

In phase 6 the archiving of the updated risk register is required. All residual and unresolved risks are to be transferred to the operations risk register. The final turnaround audit takes place and recommendations are included in the turnaround closeout report.

8.8.1 Post-turnaround audit

This takes place prior to the finalization of the turnaround closeout report and the findings are included in this report.

8.8.2 Risk records and transfer of remaining risks

The risk register is updated and outstanding risks are transferred to the company risk system.

A final risk report is included in the lessons learned/turnaround report.

Final HSE statistics are produced.

All relevant statistics for benchmarking are collated and analyzed relative to industry best practices.

KPI values and targets are published.

8.9 Procurement
8.9.0 Introduction

In phase 6 turnaround execution contracts are wound down. All contractors are evaluated and contractual documents completed.

8.9.1 Contractor and supplier evaluations

Each contractor requires a formal written evaluation for reference in the turnaround report. The supervisor responsible for the contractor is probably best suited to prepare this. Items to be included in this evaluation should include:

1. Quality, experience, and training of supervisors
2. Experience and training of artisans
3. Professionalism with regard to contractual obligations
4. Familiarity with the plant

5. HSE awareness and performance
6. Overall contribution to the turnaround success
7. Potential for partnering (if this is not a partnered contractor)
8. ISO 9000 quality certification
9. Compliance with quality standards and quality of work.

Equipment and materials suppliers also need to be evaluated. Their evaluations could be entered in the enterprise resource management system materials module. Details for specific turnarounds could be extracted for filing, and also referenced in the turnaround report. These reports could include the following:

1. Actual delivery date versus promised delivery date
2. Quality problems on delivery
3. Packaging problems
4. Shipping complications
5. Variations to contract
6. ISO 9000 quality certification.

Feedback from the contractors and suppliers is essential to determine and document any concerns experienced during the turnaround. Their recommendations for improvement are also vital and must be documented.

8.9.2 Completion of contractual documentation

All variation orders are to be agreed on and finally approved prior to arriving at final contract values. Payment is made and contracts closed out.

Capital work carried out during the turnaround would be charged to separate accounts. Settlement of these accounts may take longer than expected as items such as retention would have to be accounted for.

8.10 Stakeholder management
8.10.0 Introduction and relevant PM processes

In phase 6 final reporting to the relevant stakeholders takes place. A note expressing appreciation of their support throughout the turnaround stands one in good stead for the next turnaround.

Related standard processes are:

PMBOK	ISO 21500
13.3 Manage stakeholder engagement 13.4 Monitor stakeholder engagement	4.3.10 Manage stakeholders

See Appendix A10 for details.

8.10.1 Final discussions and project closure

Besides presenting the final turnaround report to the decision executive and other key stakeholders, details of lessons learned need to be openly discussed to ensure agreement on all improvements that will be applied in the next turnaround.

8.11 Summary: phase 6—post-turnaround
8.11.1 Focus areas

1. Demobilize contractors
2. Post-turnaround cleanup
3. Disposal of excess material
4. As-built drawings
5. Repair and inspection history reports
6. Equipment database update
7. Lessons learned and recommendations for future turnarounds
8. Compare work done with work planned
9. Outstanding work for onstream maintenance
10. Freeze turnaround accounts
11. Issue final cost report
12. Issue final turnaround report

8.11.2 Deliverables

1. Post-turnaround cleanup
2. Final reports
3. Final audit

8.11.3 Tools and processes

Final turnaround report.

8.11.4 Timing

One month after turnaround.

Reference

1. BP Texas City Refinery (turnaround report format).

Further reading

Singh R. *World class turnaround management*. Everest Press; 2000.

Conclusion

Conclusion

CHAPTER

Getting to pacesetter

9

9.0 Introduction

Turnarounds relate directly to the profitability of the company. In order to achieve higher profitability, the company needs to move toward being a pacesetter. This requires a clear company commitment and strong leadership support to change the culture of the company.

9.1 Definitions of a pacesetter
Definition: generic

An organization that is the most progressive or successful and serves as a model to be imitated.

Definition: PTAI[1]

The average of two sites nearest the top of the comparison group.

Definition: Solomon Associates[2]

Achieve first- or second-quartile accomplishments in all the major performance areas for the entire trend period (three consecutive studies − 6 years).

9.2 Alternative approaches

Physical assets and people are put at risk if the company focus is on reduction of costs without first optimizing utilization. As noted in Chapter 1, utilization relates to production and revenue generation, and is supported by availability, which in turn is determined by planned and unplanned down days. **The optimum down days are the result of well-planned and executed turnarounds.** If arbitrary cost reduction is imposed, disaster ensues. Box 9.1 illustrates British Petroleum's track record as a result of arbitrary cost reduction.

Fig. 9.1 shows both the correct and incorrect routes to achieving pacesetter status. Improving utilization is imperative before maintenance costs can be reduced. Maintenance costs typically reduce with improved utilization as a result of an improvement in reliability.

> **Box 9.1: The battering of British Petroleum's reputation[3]**
>
> After a series of incidents around the world, British Petroleum's reputation has taken a hard knock, the biggest being the following:
> - *Grangemouth Refinery—explosion in 2000*
> - *Texas City Refinery—explosion in 2005*
> - *Alaska Pipeline—leaks in 2010 and 2011*
> - *Exploration in the Gulf of Mexico (Deepwater Horizon)—explosion and leak in 2010*

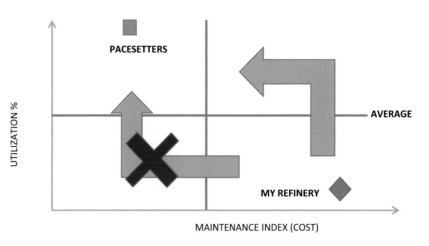

FIGURE 9.1

Routes to achieving pacesetter status.[4]

9.3 Improvement process

> Good factual information is required to make good decisions.

During the early stages of an improvement drive there will be few, if any, benchmarking results with which to compare. Facts are needed for submission for benchmarking purposes to establish a baseline from which to work. So where to start when these are not available? An enlightened leadership team will encourage open discussion to identify where the problems lie. Different techniques such as Dr. Edward de Bono's "6 Hats,"[5] or brainstorming, could be used to good effect to elicit information.

The basic process is:

1. Identify, analyze, and evaluate major gaps
2. Set performance targets
3. Agree on an improvement plan
4. Implement the plan.

1. Identify, analyze, and evaluate major gaps

The collation of necessary information for assessment is required before undertaking a structured analysis of issues to be resolved. The assessment identifies gaps, and analyzes, evaluates, and prioritizes them. This structured analysis is normally undertaken by an experienced multidisciplinary team.

Some questions to be asked:

Specific turnaround

1) Was the duration of the last turnaround perceived to be excessive?
This relates both to in-company performance and in comparison to other companies with the same plant (if benchmarking has been done).
2) Was there a cost overrun?
If so, either the budget was inadequate or there were excessive requests for additional work.
3) Were there a number of health, safety, or environmental (HSE) incidents?
Was HSE enforcement and individual safety awareness lacking?
4) Was the start-up delayed?
What caused the delay? Items could include leaks, spades not removed, work not completed, etc.
5) Were there many additional work requests?
What were they related to? This would include discovery of additional work when opening up the plant, or other work requested by various stakeholders, etc.
6) Were there work stoppages or delays in getting access to the workplace?
Items such as delays in getting work permits, availability of resources, etc. need to be reviewed.
7) Were there communication issues that inhibited the progress of work?
An example could be contractors working too close to each other, causing one contractor to stop work for safety reasons.

General

8) Is total maintenance cost perceived to be excessive (onstream and turnaround)?
The company may be spending more than the norm while it is enhancing its reliability by investing in asset management software, or it may just be inefficient.
9) What is the run length between turnarounds? Is this in line with industry norms?
The amount of unplanned downtime may be an indicator of excessive run length, or just poor onstream maintenance and inadequate turnaround work.
10) Is there a structured, phased approach to managing turnarounds?
There need to be clear phases with agreement that the work that is to be done in each phase is completed before continuing to the next phase.
Example of major gap: identify, analyze, and evaluate.
Duration of the last crude distillation unit turnaround was 40 days, which is well in excess of industry norms. The duration of typical crude distillation unit turnarounds is 28 days (obtained from benchmarking data).

2. Set performance targets

Targets can be derived from benchmarking or internal estimates of what can be achieved.

Example of major gap: set targets.

Target for the next turnaround is 34 days and the long-term target is 28 days.

3. Agree on an improvement plan

Decision and actioning parties need to agree on action plans and the time frame in which to implement them. The plans should identify only those major initiatives that need to be addressed. Key activities are:

a) Assign a sponsor to oversee the implementation plans
b) Prioritize recommendations and action items
c) Assign individual team leaders to each recommendation to ensure timely completion
d) Establish time frames for completion
e) Conduct monthly reviews of progress to ensure recommendations are being implemented
f) Embed the solutions into the turnaround processes.

Example of major gap: improvement plan to achieve targets.

Plan outline:

i. *Establish a turnaround improvement committee that reports to the CEO (sponsor).*
ii. *Review the company tendering policy for incentive-based turnaround contracts.*
iii. *Appoint a turnaround project manager for the next turnaround.*
iv. *Apply a structured scope review process for the next turnaround.*
v. *Invest in critical path software and training of full-time planning staff.*
vi. *Draft/update the turnaround framework and manual.*

4. Implement the plans

It should be noted that there is often resistance to change. The application of Kotter's eight steps[6] could be used to manage this and to ensure implementation of action plans.

9.4 Benchmarking

Benchmarking is essential to ensure continuous improvement.

Operational benchmarking is common for refineries and gas plants. Refineries normally use Solomon[2] Associates, whereas gas plants tend to use PTAI.[1] The turnaround data is often embedded in the benchmarking reports and has to be extracted on a piecemeal basis. Data includes duration, run length, utilization, availability, reliability, turnaround cost, onstream maintenance cost, etc.

IPA[7] does turnaround-specific benchmarking for the process industry, which focuses on four primary indicators: schedule efficiency (turnaround duration relative to others with similar scope and complexity), cost efficiency (cost relative to others with similar scope and complexity), availability (outside turnaround periods), and safety, health, environment, and quality (SHEQ) incident rates during turnarounds. The driving forces for improving these four primary indicators are:

1. Project management practices
2. Project management tools and processes (procedures)
3. Maintenance and inspection tools.

These are explained further:

1. Project management practices

a. Use defined turnaround management processes
b. Align turnaround scope and objectives with the business objectives
c. Align contracting strategy with the turnaround objectives
d. Manage the turnaround like a project using a dedicated team with defined roles and responsibilities
e. Enlist all business functions as part of the turnaround process
f. Use effective scoping, costing, and planning methods
g. Optimize the pre-turnaround work
h. Achieve best practiced front-end loading and freeze scope early.

2. Project management tools and processes (procedures)

i. Build a planning and costing model using high-level planning software linked to the corporate accounting, maintenance management, and time-keeping systems, and write a procedure for proper model establishment and management
j. Create standard planning templates from previous turnarounds for inclusion in the above model
k. Establish a change control procedure to be strictly applied after the scope freeze date when additional work is requested
l. Establish quality control procedures to be strictly applied during the execution phase
m. Establish an incident management procedure to ensure all lost time and high-value incidents in SHEQ areas are recorded
n. Establish a contract administration procedure to ensure full cost control, especially during the execution phase.

3. Maintenance and inspection tools

o. Establish standard spares templates on the maintenance management system from previous turnarounds for the purchase of spares required for the turnaround
p. Use decision tools for inclusion or exclusion from the scope of the turnaround
q. Use alternative practices to reduce turnaround time.

The relationships between practices (driving forces) and outcomes (indicators) are shown in Fig. 9.2.

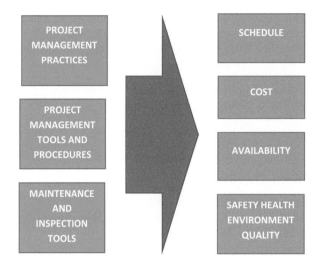

FIGURE 9.2

Relationships.[8]

Box 9.2: Marathon petroleum company (MPC)[9]

MPC's turnaround performance is competitive in the industry. Our overall company performance was in the second quartile of the 2014 Solomon Fuels Study, with three of our refineries performing in the first quartile. ..., we couldn't be more pleased with our first-quarter turnaround performance in 2017.

See also Section 1.3.2: Example of high utilization[10] and Box C3: Seven refineries, one solution.[11]

Extract from Marathon News.

An example of a pacesetter refinery is given in Box 9.2.

Benchmarking only tells you where the company is with regard to turnaround management. Analysis of the causes of the gaps and a motivation to improve are required to get to pacesetter.

9.5 Conclusion

The fundamental phases required for pacesetter turnaround management are summarized in Fig. 9.3.

With good leadership skills on the part of the turnaround project manager, a structured scope challenge, suitable computerized asset management and critical path tools, as well as motivated contractors, the consistent application of these phases will go a long way toward achieving pacesetter status.

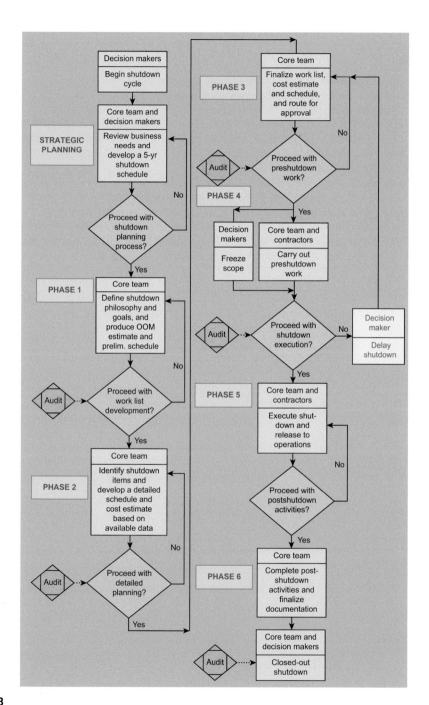

FIGURE 9.3

Process flow diagram showing basics of each phase.[8]

References

1. Phillip Townsend Associates (PTAI). Available from: http://www.ptai.com/.
2. Solomon Associates. Available from: https://www.solomononline.com/.
3. Steffy LC. *Drowning in oil, BP and the reckless pursuit of profit*. McGraw-Hill; 2011.
4. Hey RB. *Performance management for the oil, gas and process industries*. Elsevier; 2017. Available from: https://www.elsevier.com/books/performance-management-for-the-oil-gas-and-process-industries/hey/978-0-12-810446-0.
5. De Bono E. *Six thinking hats*. Penguin; 2000.
6. Kotter J, Rathgeber H. *Our iceberg is melting*. Macmillan; 2006.
7. Independent Project Analysis (IPA). Available from: http://www.ipaglobal.com/.
8. Hey RB. Make your next turnaround the best ever. *Hydrocarbon Processing* April 2001.
9. Marathon Petroleum Company News; 2017. Available from: https://news.marathonpetroleum.com/.
10. Crowley K. *Big oil leaves analysts fuming about being in the dark on refinery outages*. Bloomberg; 2018.
11. Kreiling J. Seven refineries, one solution. Primavera Magazine. Available from: www.primavera.com.

Appendices

Appendices

Appendix A: Project management standards

A0. Overview

The commonly accepted guides and standards for project management are *A Guide to the Project Management Body of Knowledge (PMBOK Guide),* including *The Standard For Project Management (ANSI/PMI 99-001-2017),* and *Guidance on Project Management ISO 21500:2012.* The former is commonly referred to as PMBOK. Although both are discussed in this appendix, only the titles described in PMBOK will be used to outline each process.

Notes:
- *Items in italics are comments specific to turnarounds*
- ***Items in bold italics are of special interest for turnarounds***
- *For the sake of simplicity, the processes described here are from start to finish of the turnaround, and not each phase.*

Table A1 shows project management process groups and knowledge areas related to turnaround phases.

Fig. A1 shows a simplified version of the table related to typical software used for project management.

A1. Integration management
A1.0 Introduction

Integration management includes the processes and activities needed to identify, define, combine, unify, and coordinate the various processes related to the project.

The integrative project management processes include:

PMBOK	ISO 21500
4.1. Develop project charter	4.3.2 Develop project charter
4.2. Develop project management plan	4.3.3 Develop project plans
4.3. Direct and manage project work	4.3.4 Direct project work
4.5. Monitor and control project work	4.3.5 Control project work
4.6. Perform integrated change control	4.3.6 Control changes
4.7. Close project or phase	4.3.7 Close project phase or project
4.4. Manage project knowledge	4.3.8 Collect lessons learned

		Project management process groups (applied to the complete project)		
	Knowledge areas	**Initiating**	**Planning**	
Ref no	**Turnaround phases**	**1. Conceptual development**	**2. Work development**	**3. Detailed planning**
1	4. Project integration management	4.1 Develop project charter 4.2a Develop project management plan	4.2b Develop project management plan	4.2c Develop project management plan
2	5. Project scope management	5.1a Plan scope management	5.1b Plan scope management 5.2 Collect requirements 5.3 Define scope	5.4 Create WBS 5.5a Validate scope
3	6. Project schedule management	6.1a Plan schedule management 6.2a Define activities	6.1b Plan schedule management 6.2b Define activities	6.3 Sequence activities 6.4 Estimate activity durations 6.5 Develop schedule
4	7. Project cost management	7.1a Plan cost management 7.2a Estimate costs (OOM) 7.3a Determine budget (BP&B)	7.1b Plan cost management 7.2b Estimate costs 7.3b Determine budget (prelim)	7.2c Estimate costs 7.3c Determine budget (final)
5	8. Project quality management	8.1a Plan quality management	8.1b Plan quality management	8.1c Plan quality management
6	9. Project resource management	9.1a Plan resource management 9.2a Estimate activity resources (generic) 9.3a Acquire resources	9.1b Plan resource management 9.2b Estimate activity resources (generic) 9.4a Develop project team (core)	9.1c Plan resource management 9.2c Estimate activity resources (specific) 9.3b Acquire resources (contractors)
7	10. Project Communica-tions management	10.1a Plan Comm management	10.1b Plan Comm management	10.1c Plan Comm management
8	11. Project risk management	11.1a Plan risk management	11.1b Plan risk management	11.2 Identify risks 11.3 perform qualitative risk analysis 11.4 Perform quantitative risk analysis 11.5 Plan risk responses
9	12. Project procurement management	12.1a Plan procurement management	12.1b Plan procurement management	12.1c Plan procurement management 12.2a Conduct procurements
10	13. Project stakeholder management	13.1 Identify stakeholders 13.2a Plan stakeholder engagement	13.3a Manage stakeholder engagement	13.3b Manage stakeholder engagement

Table A1 Project management process groups and knowledge areas.

Adapted from PMBOK tables 1—4.

Table A1 Project management process groups and knowledge areas.—cont'd			
Project management process groups (applied to the complete project)			
Executing		**Monitoring & controlling**	**Closing**
4. Pre-turnaround	**5. Turnaround execution**		**6. Post-turnaround**
4.3a Direct & manage project work	4.3b Direct & manage project work	4.5 Monitor & control project work 4.6 Perform integrated change control 5.5b Validate scope 5.6 Control scope	4.4 Manage project knowledge 4.7 Close project
		6.6 Control schedule	
		7.4 Control costs	
8.2a Manage quality	8.2b Manage quality	8.3 Control quality	
9.3c Acquire project team (contractors) 9.4b Develop project team (all)	9.5 Manage project team	9.6 Control resources	
10.2a Manage communication	10.2b Manage communication	10.3 Monitor communication	
11.6a Implement risk responses	11.6b Implement risk responses	11.7 Monitor risks	
12.2b Conduct procurements	12.2c Conduct procurements	12.3 Monitor procurements	
13.3c Manage stakeholder engagement	13.3d Manage stakeholder engagement	13.4 Monitor stakeholder engagement	

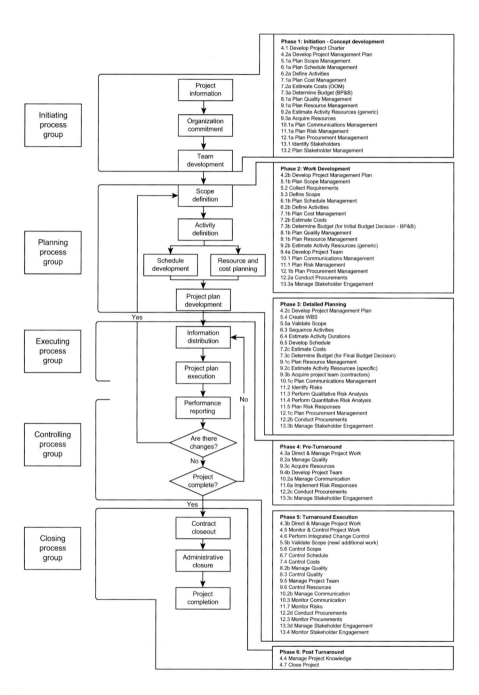

FIGURE A1

Process flow for project management software.

A1.1 **Develop project charter**

The project charter:
- Is a document that formally authorizes a project
- Provides the project manager with the authority to apply organizational resources to project activities.

The project charter is issued by a project initiator or sponsor external to the project.

Turnaround project charters are described in Section 3.1.1.

A1.2 **Develop project management plan**

The project management plan is a set of documents that describes how the project is executed, monitored, and controlled.

In the case of turnarounds, it is a summary or milestone plan in phases 1 and 2. In phase 3 it becomes a set of network diagrams in a critical path model (CPM) for execution in phases 4 and 5. *Table 3.2 shows a summary plan and Box 3.7 shows a milestone schedule.*

A1.3 **Direct and manage project work**

The project manager and the project team are required to perform actions to execute the project management plan in order to accomplish the work that has been defined in the project scope statement.

It also requires implementation of:
- Approved corrective actions—to bring anticipated project performance into compliance with the project management plan
- Approved preventative actions—to reduce the probability of potential negative consequences
- Approved defect repair requests—to correct defects found by the quality process.

Turnaround projects are managed by the turnaround project manager together with a core team (see Sections 3.6.2 and 3.6.3).

A1.4 **Manage project knowledge**

For turnarounds, this relates to the safekeeping of historical data: maintenance, inspection, operations, and turnaround records.

Lessons learned process

This focuses on identifying project successes and project failures, and includes recommendations to improve future performance on projects.

During the project life cycle, the project team and key stakeholders identify lessons learned, compile and formalize, and finally archive them.

Lessons learned studies provide future project teams with information that can increase their effectiveness and efficiency. Lessons learned sessions at the end of each phase can provide powerful team building exercises.

Results from lessons learned include:
- Update of the lessons learned knowledge base
- Input to knowledge management system
- Updated corporate policies, procedures, and processes
- Improved business skills
- Overall product and service improvements
- Updates to the risk management plan.

In turnarounds, phase end audits determine recommendations for immediate improvement, whilst the turnaround closing report records lessons learned that can be retrieved and reviewed before the next turnaround plan starts (see Sections 3.8.8, 8.7.3, and 8.7.4).

A1.5 Monitor and control project work

Project processes associated with initiating, planning, executing, and closing are carefully monitored so that corrective or preventive actions can be taken to control the project performance. Monitoring is an aspect of project management performed throughout the project.

Using Primavera or Open Plan software, monitoring and control data are entered daily to show progress and deviations from the baseline plan (see Section 7.3.1).

A1.6 Perform integrated change control

For turnarounds, scope criteria are set at the start of planning. Scope change prior to freezing of the scope is managed through a scope review process. After the scope has been frozen, a strict change control process is initiated (see Section 7.1.4*).*

A1.7 Close project or phase

This involves performing the project closure portion of the project management plan.

It includes finalizing all activities to formally close the project, and handover of the completed or canceled project as appropriate.

An effective, well-established filing system is critical as intervals between turnarounds can be in excess of 4 years with significant changes in staff. New staff must be able to access all documentation from previous turnarounds quickly and easily.

The turnaround critical path model with actual times and costs must be retained for budgeting and preparation purposes for the following turnaround.

If the turnaround is terminated prematurely, all records of both finished and unfinished work have to be retained for planning of the next turnaround.

Historical information updates after a turnaround are critical for improving the preparation and execution of the next turnaround. All inspection, maintenance, and engineering records must be updated (see Sections 8.7.2, 8.7.5 and 8.7.6).

A2. Scope management
A2.0 Introduction

Scope management includes the processes used to identify all the essential work required to be done, to successfully complete the project.

The scope management processes are:

PMBOK	ISO 21500
5.1. Plan scope management	—
5.2. Collect requirements	—
5.3. Define scope	4.3.11 Define scope
5.4. Create work breakdown structure (WBS)	4.3.12 Create work breakdown structure
See schedule	4.3.13 Define activities
5.5. Validate scope	—
5.6. Control scope	4.3.14 Control scope

A2.1 Plan scope management

The aim of scope planning is to create a scope management plan, which describes how the project team will perform each of the following:
- Collect requirements
- Define project scope and develop a detailed scope statement
- Create WBS
- Validate scope
- Control scope.

Turnaround scope planning is undertaken in two categories:
1. *Routine maintenance and inspection*
2. *Major maintenance and engineering.*

Routine maintenance and inspection relates to inspection and repair of existing physical assets.

Major maintenance and engineering relates to replacement or upgrading of existing physical assets. For example, replacement of fluid catalytic cracking unit cyclones, tie-ins for future process units, etc.

For turnarounds, work scope planning entails:
- *Setting criteria for inclusion in the turnaround (see Section 3.2.1)*
- *Selecting the participants for scope review teams (see Sections 3.6.4 and 3.6.5)*
  - *including key specialists who are required for each team*
- *Teams could be grouped according to*
  - *maintenance work orders*
  - *inspection advice tickets*
  - *project construction turnover packages—approval for expenditures*
- *Setting times and frequency of scope review meetings*
- *Deciding on tools for decision-making (see Section 4.2.2).*

Organizational culture may need to be addressed should a more "laid back" culture allow for work to be done that is not included in the approved work scope. Attitudes need to be adjusted to ensure that agreed procedures are followed. Nevertheless, all unapproved work still needs to be logged on the model.

Initial scope planning meetings require the presence of key management representatives: operations, maintenance, inspection, and engineering (see Box 3.14).

The basic rule of "If it can be done online, leave it out of the turnaround" applies when deciding what work should be included in the scope of a turnaround; Section 3.2.1 discusses scope development.

A2.2 Collect requirements

Sources of turnaround scope planning requirements include:
1. *Inspection records*
2. *Maintenance records*
3. *Plant performance records*
4. *Previous turnaround report.*

These documents are analyzed to determine what work should be included in the turnaround and are referenced in the turnaround work order. This is listed either in an automated system or in a spreadsheet, and is sometimes referred to as the preliminary work list.

A2.3 Define scope

This entails developing a detailed project scope statement as the basis for future project decisions. This is critical for the success of the project.

For turnarounds this calls for scope challenge workshops, which are discussed in detail in Sections 4.2.1, 4.2.2, and 5.2.2.

Meetings include:
- *Detailed review meetings (see Box 3.14)*
  - *Work list meeting*
  - *Major task review meeting*

- *Inspection review meeting*
- *Project work review meeting*
- *All scope*
  - *Work scope review team (see* Section 3.6.4*)*
  - *Work scope peer review by an independent peer review team (see* Sections 3.8.8 and 5.8.3*)*

A2.4 Create work breakdown structure

WBS is a deliverable-oriented hierarchical decomposition of the project work. WBS organizes and defines the total scope of the project.

For turnarounds, WBS creation is described in detail in Section 5.2.1 and Appendices C2 and C3.

A2.5 Validate scope

This is the formal acceptance by the stakeholder of the proposed and completed project scope.

In turnarounds, validation is done in two stages: the first stage is at the end of the scope challenge workshops once the scope has been frozen in the planning phase (see Section 5.2.2), and the second stage is done when the scope has been completed in the execution phase (see Section 7.3.1).

A2.6 Control scope

This process is concerned with managing those factors that create scope changes and controlling the impact of those changes. It assures all requested changes and recommended corrective actions are processed through the integrated change control process. It includes managing the changes when they occur and integrating them with other control processes.

*Any work **scope additions** or **significant scope growth** should be approved by the appropriate authorities prior to field execution.*

A clear postscope freeze procedure is required for any additional work.

Information required for decision-making/approval of additional work includes:
- *Detailed work sequence and plans*
- *Estimated work hours and resource requirements*
- *Material requirements and availability*
- *Estimated cost impact*
- *Work category priority*
- ***Schedule analysis and impact.***

The change control process for turnarounds is shown in Fig. 7.1, and an example of the process for turnarounds is shown in Box 7.2.

A3. Schedule management

A3.0 Introduction

Schedule management requires defining activities and sequencing these activities to ensure timely completion of the project.

For turnarounds, the heart of schedule management is the critical path model. This is outlined in Fig. 1.6. Appendix C covers this in detail.

Schedule management entails the following processes:

PMBOK	ISO 21500
6.1. Plan schedule management	—
6.2. Define activities	See scope
6.3. Sequence activities	4.3.21 Sequence activities
6.4. Estimate activity durations	4.3.22 Estimate activity durations
6.5. Develop schedule	4.3.23 Develop schedule
6.6. Control schedule	4.3.24 Control schedule

A3.1 Plan schedule management

For turnarounds, the output is the milestone schedule (see Section 3.3.1).

A3.2 Define activities

Previous turnaround records are used as templates for future turnarounds. Activity definition and categorization are explained in Sections 4.3.2 and 4.3.3.

A3.3 Sequence activities

See A3.5

A3.4 Estimate activity duration

See A3.5

A3.5 Develop schedule

A3.3 to A3.5 are developed in the critical path software by experienced turnaround planning engineers. Appendices C2 and C3 provide more information about the development of this computer model.

A3.6 **Control schedule**

Schedule variance (SV) and schedule performance index (SPI) are utilized as performance measurement techniques, and necessary actions are taken accordingly.

SV and SPI are built into the Primavera and Open Plan CPM software (see Appendix C7).

A4. **Cost management**
A4.0 **Introduction**

Project cost management includes processes involved in the planning, estimating, budgeting, and controlling of costs so that the project can be completed within the approved budget. It is primarily concerned with the cost of the resources needed to complete project schedule activities.

Project cost management processes include:

PMBOK	ISO 21500
7.1. Plan cost management	—
7.2. Estimate costs	4.3.25 Estimate costs
7.3. Determine budget	4.3.26 Develop budget
7.4. Control costs	4.3.27 Control costs

A4.1 **Plan cost management**

Overall cost management for all turnarounds is in alignment with the company annual planning and budgeting process. However, each turnaround project is managed separately from initiation of the project to closeout.

A4.2 **Estimate costs**
What is the cost estimating process?

It is the process of developing an approximation of the cost of the resources needed to complete each scheduled activity. In approximating cost, the estimator considers the possible causes of variation of the cost estimates, including risks.

Cost estimates are expressed in units of currency (dollars, euros, etc.), or units of measure to estimate cost, such as man-hours or man-days.

What is a schedule activity cost estimate?

It is a quantitative assessment of the likely costs of the resources required to complete the scheduled activity.

For turnarounds, cost estimating is normally carried out by experienced planning engineers who have both a thorough knowledge of what is required to be done in the turnaround as well as the ability to build a cost and schedule model in the critical path software.

A4.3 Determine budget

For turnarounds this is outlined in Section 3.4.3.

A4.4 Control costs
What is cost control?

Cost control entails:
- Influencing the factors that could change the cost baseline
- Ensuring requested changes are agreed upon
- Ensuring that potential cost overruns do not exceed the authorized funding for the project
- Monitoring cost performance to detect variances from the cost baseline
- Recording all appropriate changes accurately against the cost baseline
- Preventing incorrect, inappropriate, or unapproved changes from being included in the reported cost or resource usage
- Informing the appropriate stakeholders of approved changes
- Acting to bring expected cost overruns within acceptable limits.

Project cost control searches out the causes of positive and negative variances and is part of integrated change control. *A strict change control process for turnarounds is described in* Section 7.1.4.

Earned value management

Earned value management is a technique that objectively manages project performance and progress. It is a systematic process used to find variances in projects based on the comparison of work performed and work planned.

Specifically, the earned value technique compares the cumulative value of the budgeted cost of work performed (earned) at the original allocated budget amount to both the budgeted cost of work scheduled (planned) and to the actual cost of work performed (actual).

The basics

Schedule variance (SV). Where you are relative to where you should be at this time.

Cost variance (CV). How much you have spent relative to how much you should have spent at this time.

Earned value terms are described further in Table A2.

Earned value variance formulas are shown in Table A3.

Box A1 gives a sample scenario.

Earned value management formulas are shown in Table A4.

Table A2 Earned value management terms.

Term	Description	Interpretation
BCWS PV	Budgeted cost of work scheduled or **planned value**	The estimated value of the work planned to be done (at a certain date).
BCWP EV	Budgeted cost of work performed or **earned value**	The estimated value of the work actually done (at the same date).
ACWP AC	Actual cost of work performed or **actual cost**	The actual cost incurred (at the same date).
BAC	Budget at completion or **project budget**	The budget for the total project.

Table A3 Earned value variance formulas.

Name	Formula	Interpretation
Variance	**Plan − actual**	Positive is a good indicator, negative is a bad indicator
Cost variance (CV)	**EV − AC**	Negative is over budget, positive is under budget
Schedule variance (SV)	**EV − PV**	Negative is behind schedule, positive is ahead of schedule

Box A1 EV scenario

1. Is the project's CV negative or positive?
   - CV = EV − AC, negative
2. Is the project over or under budget?
   - **Over budget**
3. Is the project's SV negative or positive?
   - SV = EV − PV, negative
4. Is the project ahead or behind the schedule?
   - **Behind schedule.**

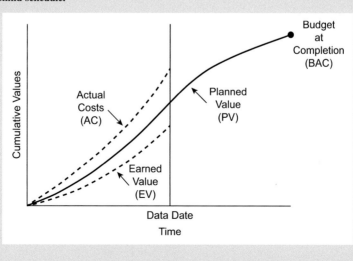

Table A4 Earned value management formulas.		
Name	**Formula**	**Interpretation**
Index	**Plan/Actual**	Value > 1 is a good indicator, value < 1 is a bad indicator
Cost performance index (CPI)	**EV/AC**	We are only getting \$X out of every \$1
Schedule performance index (SPI)	**EV/PV**	We are progressing at X% of the rate originally planned

Forecasting

This includes making estimates or predictions of possible conditions that might occur in the future of the project based on information and knowledge available at the time of the forecast.

Cumulative CPI. Cumulative cost performance index (CPI^c) is widely used to forecast project cost at completion. It equals the sum of the periodic earned values (EV^c) divided by the sum of the individual actual costs (AC^c):

$$CPI^c = EV^c/AC^c$$

Earned value management terms are shown in Table A5.

Estimate to completion. Estimate to completion (ETC) can be calculated as follows:
- Based on a new estimate: ETC equals the revised estimate for the work remaining
- Based on typical variances: (similar variances are not likely to happen in the future)
  - $ETC = BAC - EV^c$
- Based on typical variances: (Similar variances could happen in the future)
  - $ETC = (BAC - EV^c)/CPI^c$

BAC: Budget at completion.

Estimate at completion. Estimate at completion (EAC) can be calculated as follows:
- Based on a new estimate: EAC equals the actual costs to date plus a new ETC
  - $EAC = AC^c + ETC$

Table A5 Earned value management terms for completion.			
Term	**Description**	**Formula**	**Interpretation**
EAC	Estimate at completion	**BAC/CPI**	What do we currently expect the total project to cost (at certain date)?
ETC	Estimate to complete	**EAC − AC**	From this point on (at a certain date), how much more do we expect it to cost to finish the project?
VAC	Variance at completion	**BAC − EAC**	How much over or under budget do we expect to be?

- Using remaining budget: (similar variances are not likely to happen in the future)
  - $EAC = AC^c + BAC - EV^c$
- Using CPI: (similar variances could happen in the future)
  - $EAC = AC^c + ((BAC - EV^c)/CPI^c)$.

Earned value *is built into high-level CPM software such as Primavera and Open Plan (see Appendix C7).*

A5. Quality management
A5.0 Introduction

Quality management incorporates the processes required to plan and establish quality assurance and control.

The project quality management processes include:

PMBOK	ISO 21500
8.1. Plan quality management	4.3.32 Plan quality
8.2. Manage quality	4.3.33 Perform quality assurance
8.3. Control quality	4.3.34 Perform quality control

A5.1 Plan quality management

Quality planning involves identifying those quality standards that are relevant to the project and determining how to comply with them.

Standards used for the oil and gas industry related to asset and turnaround management are outlined as follows.

ISO PDCA (Plan Do Check Act) standards

The following PDCA standards are applicable to the physical assets:
- *ISO 50001 Energy Management*
- *ISO 55001 Asset Management.*

Other physical asset standards
Integrity

The following are guidelines for the integrity of the complete physical asset:
- *OGP Report No 4: Asset Integrity*
- *Center for Chemical Process Safety: Guidelines for Risk Based Process Safety.*

Inspection

The following is the basis for risk based inspection:
*   *API RP 580 Risk Based Inspection.*

Maintenance

The following are the key requirements for reliability centered maintenance:
*   *IEC 61822 Hazards & Operability Application Guide*
*   *ISO 17776 Petroleum & Natural Gas Industries—Offshore Production Installations—Guidelines On Tools & Techniques for Hazard Identification & Risk Assessment*
*   *IEC 61511 & IEC 61508 Safety Instrument Systems*
*   *SAE STD JA1011 Evaluation Criteria for Reliability Centered Maintenance Processes*
*   *ISO 14224 Petroleum, Petro Chemical & Natural Gas Industries—Collection and Exchange of Reliability and Maintenance Data for Equipment*
*   *UK HSE RR 509 Plant Aging—Management of Equipment Containing Hazardous Fluids or Pressure.*

Quality standards and codes are discussed further in Appendix B.

The application of standards generally results in quality control plans (QCPs) in a turnaround.

A5.2 Manage quality

Quality management is the application of planned, systematic quality activities to ensure that the project will employ all processes needed to meet requirements.

Adequate inspection resources are critical for managing quality in a turnaround.

A5.3 Control quality

Quality control involves monitoring specific project results to:
*   Determine whether they comply with relevant quality standards
*   Identify ways to eliminate causes of unsatisfactory results.

In turnarounds, quality is generally controlled by using QCPs. Box 5.7 gives an example.

A6. Resource management
A6.0 Introduction

Resource management includes the processes required to identify and acquire adequate project resources such as people, materials, equipment, and tools. These processes are critical to ensure that sufficient resources are timeously available to complete the project in the planned timescale.

The project resource management processes include the following:

PMBOK	ISO 21500
9.1. Plan resource management	4.3.17 Define project organization
9.2. Estimate activity resources	4.3.16 Estimate resources
9.3. Acquire resources	4.3.15 Establish project team
9.4. Develop team	4.3.18 Develop project team
9.5. Manage team	4.3.20 Manage project team
9.6. Control resources	4.3.19 Control resources

A6.1 **Plan resource management**

Plan resource management entails identifying and documenting project roles, responsibilities, and reporting relationships, as well as creating the staffing management plan and materials procurement plan.

Turnaround trades, materials, tools, and equipment resource planning are critical to ensure resources are on-site at the start of the turnaround (see Sections 3.6.7 and 3.6.8).

A6.2 **Estimate activity resources**

For turnarounds, this is carried out within the CPM and includes all resources required to complete the activity: human, financial, material, equipment, and tools (see Sections 4.6.1 and 5.6.3).

A6.3 **Acquire resources**

This entails obtaining the resources needed to complete the project. *For turnarounds, various contractors are employed for carrying out the activities. Supervision and management could be carried out either by contractors or the owner. Materials, equipment, and tool acquisition starts in phase 1 with securing long-lead items. All materials, tools, and equipment have to be on-site prior to commencement of the execution phase (see Sections 4.6.2 and 5.9.4).*

A6.4 **Develop team**

The competencies and interaction of team members are critical to enhance project performance. This involves improving the skills of team members and enhancing feelings of trust and cohesiveness to encourage productivity through teamwork.

The right people have to be selected at the start of the planning stage of the turnaround. Training of inexperienced staff and contractors and those who lack the required competencies or skills could result in a duration delay, quality rework, safety infringement, or cost increase. Experienced planners should, ideally, be trained to use the applicable software up to 18 months before execution (see Sections 3.6.5 and 3.6.6).

Generic site-specific training, such as the process for issuing permits, safety management, team building, etc., can take place in the preexecution phase (see Sections 6.6.1 and 6.6.2).

A6.5 **Manage team**

Team management involves tracking team member performance, providing feedback, resolving issues, and coordinating changes to enhance project performance. It also observes team behavior, manages conflict, and appraises team member performance.

For turnarounds, a strict hierarchical organization structure must be observed. This is published prior to the execution phase to ensure that the chain of command is very clear. The turnaround project manager should be in total control. During the execution phase, daily interaction between the turnaround project manager and his team is crucial (see Sections 7.7.3, 7.7.4, and 7.7.5).

A6.6 **Control resources**

For turnarounds, resource control is determined by work progress according to the CPM where resources have been allocated in line with the achievement of the target completion date. Daily feedback and corrective action is required to remain on schedule (see Sections 5.3.1 and 7.6.1).

A7. **Communication management**
A7.0 **Introduction**

Communication management includes the processes required to plan, manage, and distribute information relevant to the project. It ensures timely and appropriate generation, collection, distribution, storage, retrieval, and ultimate archiving of project information. Communications management is the key to project control. It provides an information lifeline among all members of the project team. Information must flow downward, upward, and laterally within the organization.

The project communications management processes include:

PMBOK	ISO 21500
10.1. Plan communications management	4.3.38 Plan communications
10.2. Manage communications	4.3.39 Distribute information
10.3. Monitor communications	4.3.40 Manage communications

A7.1 **Plan communications management**
What is communication planning?

Communication planning entails determining the information and communications needs of the project stakeholders by:
1. Ascertaining who they are
2. Determining their level of interest and influence

3. Identifying who needs what information
4. Establishing when they need it
5. Selecting how it will be given to them and by whom.

Suitable channels of communication for turnarounds:
- *Management processes and procedures*
- *Plans and strategies*
- *Goals, objectives, and performance criteria*
- *Organization structure and interfaces*
- *Coordination policies and procedures*
- *Meetings and informal discussions*
- *Correspondence, reports, and documentation*
- *E-mails, phone calls, intranet.*

A7.2 Manage communications

Communications management is a crucial element of effective turnaround management. A culture of open communication needs to be promoted to ensure appropriate information is available as needed for timeous decision-making. Particularly during the execution phase, when key decisions need to be made urgently to ensure that the schedule is maintained and that cost and quality are controlled, effective communication strategies are essential.

What is information distribution?

Information distribution involves making needed information available to project stakeholders in a timely manner. It includes implementing the communications management plan, as well as responding to unexpected requests for information.

Effective meetings

Key elements of effective meetings:
- Set and keep to a time limit
- Schedule recurring meetings in advance
- Have a purpose for the meeting
- Create an agenda with team input
- Distribute agendas beforehand
- Stick to the agenda
- Let people know their responsibilities in advance
- Bring the right people together
- Chair and lead the meeting with a set of rules
- Document and publish meeting minutes timeously.

See also Section 3.7.3.

A7.3 **Monitor communications**
What is performance reporting?

Performance reporting is the collection of all baseline and actual performance data and the distribution of performance information to stakeholders to provide them with information about how resources are being used to achieve project objectives. This includes:
- Status reporting, progress measurement, and forecasting
- Information on scope, schedule, cost, and quality, and possibly on risk and procurement.

Work performance information

Work performance information relates to the completion status of the deliverables and what has been accomplished.

Performance measurements

These are calculated CV, SV, CPI, SPI values for WBS components, and in particular, the work packages and control accounts.

Forecasted completion

These are calculated EAC and ETC values.

Quality control measurements

These are the results of QC activities that have been fed back to quality assurance staff.

Approved change requests

These are the requested and approved changes to increase or decrease the project scope and to modify the estimated approved cost and/or activity duration.

Deliverables (PMBOK definition)

A deliverable is any unique and verifiable product, result, or capability to perform a service that is required to be produced to complete a process, phase, or project.

Information presentation tools

Project performance data can be presented using different software packages.

For turnarounds, these include the critical path scheduling package and related software (see Appendices C2, C10, and C12).

Performance information gathering and compilation

Different media can be used to gather and compile the required information, e.g., manual filing systems, electronic databases, project management software, and other systems that allow access to technical documentation.

The critical path model is the key tool for turnarounds (see Appendix C2). The software for asset performance management is another conduit for gathering and compiling information (see Appendix B4).

Status review meetings

Status review meetings are regularly scheduled events to exchange information about the project, and they are held at different frequencies and levels.

It is critical to have daily progress meetings during the execution phase of turnarounds (see Section 7.7.5).

Time reporting systems

Time reporting systems (clocking systems) record and provide time expended on the project.

This is useful during the execution phase of turnarounds to determine productivity (see Sections 5.6.5 and 7.6.2).

Cost reporting systems

Cost reporting systems record and provide cost expended on the project.

It is essential to monitor costs on a daily basis during the execution phase of a turnaround. The CPM supplies key information that could be processed with information from other sources using supplementary cost management software as described in Appendix C12.

Performance reports

Performance reports organize and summarize the information gathered and present the results of any analysis as compared to the performance measurement baseline. Reports should provide the kinds of information and the level of detail required by various stakeholders as documented in the communications management plan.

Common formats of these reports include bar charts, S-curves, histograms, and tables. Earned value analysis is often included as part of these reports.

Turnaround reports are discussed further in Sections 6.7.2, 7.7.1, and Appendix C10.

Earned value analysis

Earned value analysis integrates scope, cost, and schedule measures to help the project management team assess project performance. Table A6 gives the standard terms. Detailed explanations are given in Appendix A4.4.

Forecast

*During the execution of the turnaround, information about the project's past performance determines the impact on the project in the future. Forecasts are updated and reissued **daily** based on work*

Table A6 Formulas.

Name	Formula
Cost variance (CV)	BCWP − ACWP
Schedule variance (SV)	BCWP − BCWS
Cost performance index (CPI)	BCWP/ACWP
Schedule performance index (SPI)	BCWP/BCWS
Estimate at completion (EAC)	BAC/CPI
Estimate to complete (ETC)	EAC − ACWP
Variance at completion (VAC)	BAC − EAC

performance information acquired as the project progresses. The CPM is the key tool for turnarounds (see Appendices C8, C9, and C10).

Requested changes

Analysis of project performance often generates requested changes to various aspects of the project.

For turnarounds, requested changes during the execution phase are processed in a structured manner (see Section 7.1.4).

Recommended corrective actions

Recommended corrective actions are notifications that bring the expected future performance of the project in line with the project management plan baselines.

In the execution phase of a turnaround, corrective actions are processed through the change control process (see Section 7.1.4) and QCPs (see Section 7.5.1).

Organization process assets (updates)

Updates to organization process assets include lessons learned documentation with causes of issues, reasoning behind the corrective action chosen, etc.

It is critical to update inspection and maintenance records as well as to file the turnaround model both during and at the end of a turnaround (see Sections 8.7.2−8.7.6 and 8.8.2).

A8. Risk management
A8.0 Introduction

Risk management includes the processes required to identify and manage threats and opportunities. It is the systematic process of identifying, analyzing, and responding to project risk. It includes maximizing the probability and consequences of positive events and minimizing the probability and consequences of negative events.

What is project risk?

Project risk entails dealing with an **uncertain event or condition** that will have a positive (opportunity) or a negative (threat) effect on key project objectives—scope, cost, schedule, and quality—should it occur.

Certainty relates to something that is known. For a known event, it is apparent when it will happen, how it will happen, and what the impact would be.

Uncertainty relates to an aspect of the event or the whole event being unknown.

Known—unknowns

An event that may or will happen, but it is not apparent when or how, or what the impact would be.

> *Example*
>
> *Rain may occur during a turnaround and thus cause a delay (see Section 4.8.1).*

Another example is described in Box 7.7.

Unknown—unknowns

An unforeseen event that is totally unexpected, and it cannot be known when or how it may happen or what the impact would be.

> *Example*
>
> *A tornado.*

The project risk management processes include:

PMBOK	ISO 21500	ISO 31000
11.1 Plan risk management	–	Establish the context Communication and consultation
11.2 Identify risks 11.3 Perform qualitative risk analysis 11.4 Perform quantitative risk analysis	4.3.28 Identify risks 4.3.29 Assess risks	Assessment • Risk identification • Risk analysis • Risk evaluation
11.5 Plan risk response 11.6 Implement risk responses	4.3.30 Treat risks	Risk treatment
11.7 Monitor risks	4.3.31 Control risks	Monitoring and review

A8.1 Plan risk management
What is risk management planning?

Risk management planning is deciding how to approach and plan the risk management activities for a project. The corporate framework in which risk should be managed is shown in Fig. A2.

The life of a risk is outlined in Fig. A3.

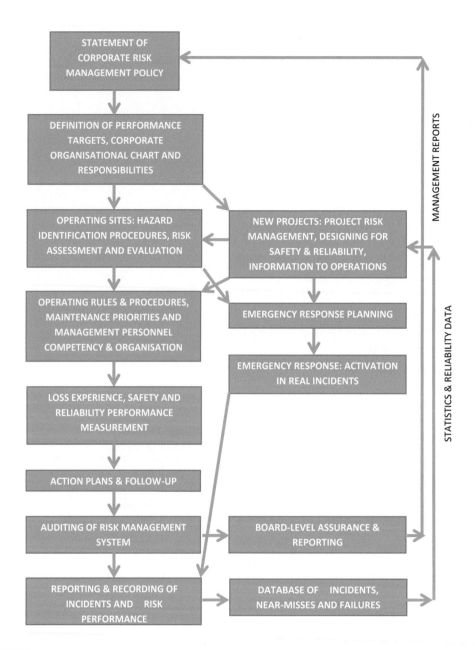

FIGURE A2

Corporate risk framework.

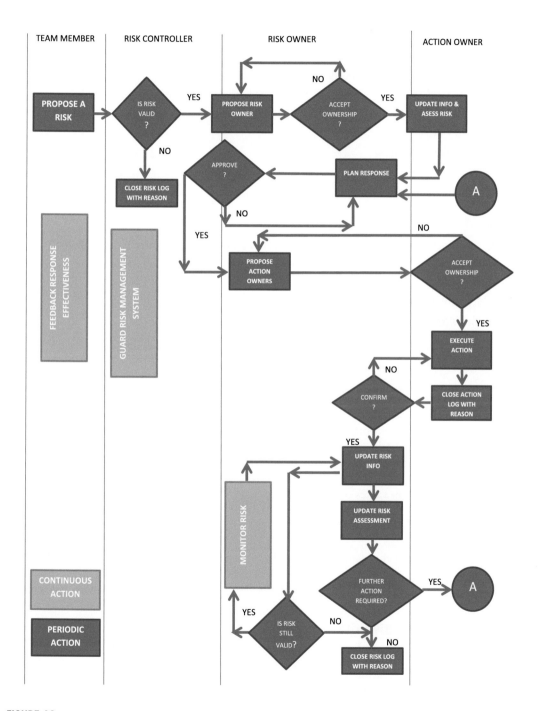

FIGURE A3

Life of a risk.

Definitions of risk probability and impact

These may be described in qualitative terms (very high, high, moderate, low, and very low).
- Risk **probability** is the likelihood that a risk will occur
- Risk **impact** (consequences) is the effect on project objectives if the risk event occurs.

A8.2 Identify risks
What is risk identification?

Risk identification entails determining which internal or external risks are likely to affect the project (positively if an opportunity and negatively if a threat), and documenting their characteristics.

Audit

Auditing often identifies potential risks for attention. Audits should be carried out at the end of each phase of the project. *Planned turnaround audits are listed in* Section 3.8.8.

A8.3 Perform qualitative risk analysis
What is qualitative risk analysis?

Qualitative risk analysis is the process of assessing the impact and probability of identified risks. A probability and impact matrix is a useful tool.

Probability/impact matrix

A matrix may be constructed that assigns risk ratings (very high, high, moderate, low, and very low) to risks or conditions based on combining probability and impact scales.
- *Risk's probability scale* falls between 0.0 (no probability) and 1.0 (certainty).
- **Risk's impact scale** reflects the severity of its effect on the project objective. Impact can be:
  - Ordinal scales: very low, low, moderate, high, very high
  - Cardinal scale: assign values to these impacts:
    - linear (0.1/0.3/0.5/0.7/0.9) or
    - nonlinear (0.05/0.1/0.2/0.4/0.8)

Box A2 shows an example of a qualitative risk matrix. Each risk is rated on its probability of occurring and the impact if it does occur. The organization's thresholds for low (green, 0.01—0.15), moderate (yellow, 0.21—0.49), or high (red, 0.63—0.81) risk as shown in the matrix determines the risk's score.

A8.4 Perform quantitative risk analysis
What is quantitative risk analysis?

Quantitative risk analysis entails numerically measuring the probability and consequences (impact) of risks, and estimating their implication for project objectives as well as the extent of the risk to the overall project.

Box A2 Qualitative risk matrix example

Impact Probability	0.9	0.7	0.5	0.3	0.1
0.9	0.81	0.63	0.45	0.27	0.09
0.7	0.63	0.49	0.35	0.21	0.07
0.5	0.45	0.35	0.25	0.15	0.05
0.3	0.27	0.21	0.15	0.09	0.03
0.1	0.09	0.07	0.05	0.03	0.01

Defined Conditions for Impact Scales of a Risk on Major Project Objectives
(Examples are shown for negative impacts only)

Project Objective	Relative or numerical scales are shown				
	Very low /.05	Low /.10	Moderate /.20	High /.40	Very high /.80
Cost	Insignificant cost increase	<10% cost increase	10-20% cost increase	20-40% cost increase	>40% cost lincrease
Time	Insignificant time increase	<5% time increase	5-10% time increase	10-20% time ilncrease	>20% time increase
Scope	Scope change barely noticeable	Minor areas of scope affected	Major areas of scope affected	Scope change unacceptable	Project end item is effectively useless
Quality	Quality degradation barely noticeable	Only very demanding applications are affected	Quality reduction requires sponsor approval	Quality reduction unacceptable to sponsor	Project end item is effectively useless

This table presents examples of risk impact definitions for four different project objectives. They should be tailored in the Risk Management Planning process to the individual project and to the organization's risk thresholds. Impact definitions can be developed for opportunities in a similar way.

This process uses techniques such as ***Monte Carlo*** simulation and decision analysis to:
- Determine the probability of achieving a specific project objective.
  Example
  ○ *There is only an 80% chance of completing the project within the 6 months required by the customer.*
  ○ *There is only a 75% chance of completing the project within the $8 million budget.*
- Quantify the risk exposure for the project, and determine the size of cost and schedule contingency reserves that may be needed.
- Identify the risks that require the most attention by quantifying their relative contribution to project risk.
- Identify achievable and realistic cost, schedule, or scope targets.

Examples of cost, duration, and completion date probabilities are shown in Appendix C9.

Event potential matrix

The event potential matrix (EPM), which consists of a risk assessment matrix (RAM) and a risk value matrix (RVM), is a very useful tool that is most commonly used to assess how important corrective action is.

An EPM is normally used to determine both the level of involvement as well as the priority/urgency of remedial action.

- **Risk Value = Probability x Rating**
- **Probability** = Likelihood of potential consequence happening
- **Rating (severity)** = Cost or effect of the **consequence** (actual or potential)

Consequences are classified as follows:
1. People: Risk to safety or health
2. Asset/Production: Risk to assets or production
3. Environment: Risk to the environment
4. Reputation/Customer: Risk to the company's reputation or customer satisfaction.

Consequence values are assigned within five levels of severity from low and high.

Probability (likelihood) is categorized from A (never heard of in industry) to E (occurs several times a year in this plant).

A rating (severity) or class is given from one to five for minor to catastrophic consequence.

Risk tolerance (low, medium, high, or intolerable) is determined from both how often the incident has occurred, as well as the rating.

The EPM is depicted in Box A3.

A8.5 **Plan risk responses**
What is risk response planning?

Risk response planning entails developing options and determining actions to enhance opportunities and reduce threats to the project's objectives.

Strategies for negative risks or threats
- Avoid (**Terminate**): Eliminate a specific threat, usually by eliminating the cause
- Transfer: Shift the consequences of a risk, together with ownership of the response, to a third party
- Mitigate (**Treat**): Reduce the expected monetary value of a risk event by reducing the probability of occurrence, reducing the risk event value, or both, to an acceptable threshold.

Strategies for positive risks or opportunities
- Exploit: This strategy seeks to eliminate the uncertainty associated with a particular upside risk by making the opportunity definitely happen
- Share: Sharing a positive risk involves allocating ownership to a third party who is best enabled to capture the opportunity for the benefit of the project

Box A3 Event potential matrix (EVM) example

EVENT POTENTIAL MATRIX
EXPANDED VERSION FOR REFINERY

RISK ASSESSMENT MATRIX (RAM)				RISK VALUE MATRIX (RVM)					
						PROBABILITY			
CONSEQUENCES					LOW <=	**LIKELIHOOD** (OF POTENTIAL CONSEQUENCE OCCURRING)		=> HIGH	
PEOPLE (RISK TO SAFETY / HEALTH)	ASSET / PRODUCTION (FINANCIAL RISK)	ENVIRONMENT (RISK TO ENVIRONMENT)	REPUTATION / CUSTOMER (RISK TO CUSTOMER OR REPUTATION)	NEVER HEARD OF IN INDUSTRY (IMPROBABLE)	HAS OCCURRED IN INDUSTRY (POSSIBLE)	HAS OCCURRED IN COMPANY	OCCURS SEVERAL TIMES A YEAR IN COMPANY	OCCURS SEVERAL TIMES A YEAR IN REFINERY	*RATING / CLASS*
Pe	As	En	Cu	A	B	C	D	E	
SLIGHT HEALTH EFFECT / INJURY (FIRST AID)	SLIGHT DAMAGE (<MR 25 000), NO DISRUPTION TO OPERATION	SLIGHT EFFECT	INTERNAL QUALITY ISSUE, WAIVER, SLIGHT IMPACT	LOW					1
MINOR HEALTH EFFECT / INJURY (LOST TIME INJURY)	MINOR DAMAGE (MR 25 000 to MR 250 000)	MINOR EFFECT	FAILURE TO SUPPLY, DELIVERY OFF SPEC, LIMITED IMPACT		MEDIUM				2
MAJOR HEALTH EFFECT / INJURY	LOCAL DAMAGE (MR 250 000 to MR 1.25 million)	LOCALIZED EFFECT (IN REFINERY)	MULTIPLE COMPLAINTS, NATIONAL IMPACT			HIGH			3
SINGLE FATALITY OR PERMANENT TOTAL DISABILITY	MAJOR DAMAGE (MR 1.25 million to MR 25 million)	MAJOR EFFECT (IN INDUSTRIAL CITY)	LOSS OF CUSTOMER/S, REGIONAL IMPACT				INTOLERABLE		4
MULTIPLE FATALITIES	EXTENSIVE DAMAGE (>MR 25 million)	MASSIVE EFFECT IN AREA / COUNTRY	LOSS OF MARKET SHARE, INTERNATIONAL IMPACT						5

SEVERITY — *=> LOW*, *ACTUAL OR POTENTIAL CONSEQUENCE*, *HIGH <=*

Risk Assessment used for Incidents that have occurred in Refinery **Risk Value & Risk Assessment** used for determining the Risk Value for potential risks

Example: spiral heat exchanger over pressure
- *Contractor overpressured spiral on hydro-test (did not use restraining clamps as per pressure test procedure)*
- *Spiral replacement cost in excess of Malaysian Ringits MR 100,000*
- *In white range of RAM with rating 2 and no previous occurrence in company but has occurred in industry; therefore, medium risk*
- *Risk to be managed within turnaround risk management system.*

- Enhance: This strategy modifies the "size" of an opportunity by increasing probability and/or positive impacts, and by identifying and maximizing key drives of these positive impact risks.

Strategies for both threats and opportunities
- Acceptance (Tolerate): Do not change project plan, but develop a contingency plan (positive action), should the risk event occur. Passive action might mean accepting a lower profit if some activities overrun.

Contingent response strategy

For some risks, it is appropriate for the project team to make a response plan that will only be executed under certain predefined conditions, such as missing intermediate milestones.

Negative risk response alternatives are summarized as **tolerate, treat, transfer, or terminate**.

Fig. A4 shows the classic 4 T's.

Risk register

Risk registers should include:
- Identified risks and descriptions
- Risk owners and assigned responsibilities
- Results from the qualitative and quantitative risk analysis
- Agreed responses including avoidance, transference, mitigation, or acceptance for each risk in the risk response plan
- Residual risks that are expected to remain after planned responses have been taken, as well as those that have been deliberately accepted

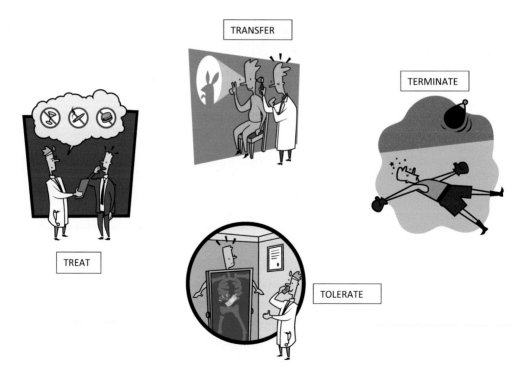

FIGURE A4

Risk response alternatives.

- Specific actions to implement the chosen response strategy
- Budget and times for responses
- Contingency plans and triggers that call for their execution
- Fallback plans for use as a reaction to a risk that has occurred, where the primary response proves to be inadequate.

Risk registers for turnarounds are discussed in Section 5.8.2.

Residual risks

Residual risks are those that remain after avoidance, transfer, or mitigation responses have been taken.

Secondary risks

Risks that arise as a direct result of implementing a risk response are termed *secondary risks*.

> *Example:*
> *The risk of a fire could be shifted to an insurance company who would bear the financial burden should a fire occur. However, a fire might cause the plant to stop production, resulting in lost profit, which might not have been insured against.*

Contingency reserves

Contingency reserves of time and cost are designed to provide for stakeholders' risk tolerances.

Contingency reserves could be calculated based on the quantitative analysis of the project and the organization's risk thresholds. *Turnaround contingencies are discussed in* Sections 5.3.2 and 5.4.3.

A8.6 Implement risk responses

Risk response implementation entails actioning agreed mitigation steps by specific actioners within an agreed time frame.

A8.7 Monitor risks
What is risk monitoring?

Risk monitoring entails:
- Keeping track of the identified risks
- Monitoring residual risks
- Identifying new risks
- Ensuring the execution of risk plans
- Evaluating their effectiveness in reducing risk.

A9. Procurement management
A9.0 Introduction

Project procurement management includes the processes required to plan and purchase or acquire product, services or results, in order to achieve project scope, from outside the performing organization.

The procurement management processes include:

PMBOK	ISO 21500
12.1. Plan procurement management	4.3.35 Plan procurements
12.2. Conduct procurements	4.3.36 Select suppliers
12.3. Control procurements	4.3.37 Administer procurements

A9.1 Plan procurement management
Buyer-seller relationship

Buyer: The customer.

Seller: Contractor; vendor; supplier external to the performing organization.

T's & C's: (terms and conditions) of the contract become a key input to many of the seller's processes.

What is procurement planning?

Procurement planning entails determining which project needs can best be met by purchasing or acquiring products, services, or results outside the organization and should be accomplished during the project planning phase. It involves consideration of whether, how, what and how much to procure, as well as when to acquire.

Extensive procurement planning takes place for a turnaround with the objective of minimizing the duration and cost, and optimizing the quality of workmanship (see Sections 3.9.1 and 4.9.1). Contracting strategy is discussed in Sections 3.9.4 and 3.9.5.

A9.2 Conduct procurements
Select sellers

The "select sellers" process involves receiving bids or proposals and applying evaluation criteria, as applicable, to select one or more vendors who are qualified and acceptable as sellers. It includes obtaining responses (*proposals and bids*) from prospective sellers on how project needs can best be met. Most of the actual effort in this process is expended by the prospective vendors, normally at no cost to the project.

Turnaround tendering and award is discussed in Sections 5.9.3 and 5.9.4.

A9.3 Control procurements
What is contract administration?

Contract administration ensures that the seller's performance meets contractual requirements.

Due to the legal nature of the contractual relationship, it is crucial that the project team be acutely aware of the legal suggestions of actions they take during administration of any contract.

Payment terms should be defined within the contract.

Contract administration includes application of the appropriate project management processes to the contractual relationship/s. The project management processes that must be applied include:

- **Conduct and monitor procurements:** (PMBOK Sections 12.2 and 12.3) to ensure proper overall management of the contracts
- **Direct and manage project work:** (PMBOK Section 4.3) to authorize the contractor's work at the appropriate time
- **Manage and monitor communication—performance reporting:** (PMBOK Sections 10.2 and 10.3) to monitor contractor cost, schedule, and technical performance
- **Manage and control quality:** (PMBOK Sections 8.2 and 8.3) to inspect and verify the adequacy of the contractor's product
- **Perform integrated change control:** (PMBOK Section 4.6) to assure that changes are properly approved, and that all those with a need to know are aware of such changes
- **Implement risk responses and monitor risks:** (PMBOK Sections 11.6 and 11.7) to ensure that risks are mitigated.

Turnaround contract management is discussed in Section 7.9.1.

A10. Stakeholder management
A10.0 Introduction

Stakeholder management includes the processes required to identify and manage the project sponsor, customers, and other stakeholders.

The stakeholder management processes include:

PMBOK	ISO 21500
13.1. Identify stakeholders	4.3.9 Identify stakeholders
13.2. Plan stakeholder engagement	—
13.3. Manage stakeholder engagement	4.3.10 Manage stakeholders
13.4 Monitor stakeholder engagement	—

A10.1 Identify stakeholders

For turnarounds, it is critical to establish all stakeholders in phase 1 (see Section 3.10.1).

A10.2 Plan stakeholder management

For a turnaround, stakeholder roles need to be clarified before the stakeholders can be managed. Fig. 3.5 shows stakeholder relationships.

Table A7 Issue log.								
Issue no	Logging date	Description	Status	Owner	Target date	Date closed	Cost impact	Comments

A10.3 Manage stakeholder engagement

Properly involving and managing stakeholders is a critical success factor (see Section 3.10.2).

A10.4 Monitor stakeholder engagement

The decision executive is a key player in turnaround projects and thus needs to be appraised regularly of progress. Close coordination with key stakeholders is particularly crucial at start-up (see Sections 7.10.1 and 7.10.2). The decision executive is presented with the final turnaround report at the end of the turnaround (see Section 8.10.1).

Issue log

An issue log is a tool that can be used to document and monitor the resolution of issues. Each issue is listed with an owner and a target date set for resolution. Table A7 shows a possible format.

Further reading

1. Guide to Project Management Body Of Knowledge PMBOK (ANSI/PMI 99-001-2017).
2. Guidance on project management ISO 21500:2012.
3. Fleming Q, Koppelman J. *Earned value project management*. PMI; 1996.
4. Ertl B. Applying PMBOK to shutdowns, turnarounds and outages. *Maintenance and Asset Management Journal* 2005;**20**(3).

Appendix B: Physical asset performance management

B0. Overview

In the process industry, the life of the physical asset is normally in excess of 30 years.

An optimum level of physical asset maintenance and reinvestment is required to ensure attainment of maximum production levels over its lifetime, and thus maximize profits. Human and financial assets, with the support of information, make sure this happens. However, if maintenance activity and costs are reduced prior to obtaining a high level of reliability, maximum long-term production levels will not be achieved.

B1. Application of asset performance management
Primary focus

Primary focus areas are:
1. Corporate safety **culture**
2. Process safety/integrity management **systems**
3. **Performance** evaluation, corrective action, and corporate oversight.

Key elements

Key elements of physical asset performance management:
a) Reliability centered maintenance (RCM)
b) Condition-based monitoring (CBM)
c) Safety instrument function (SIF)
d) Riskbased inspection (RBI)
e) Turnaround optimization
f) Initiative activity ranking
g) Priority setting of corrective maintenance
h) Defect elimination

RCM, CBM, SIF, and RBI are discussed in the following paragraphs.

Initiative activity ranking, priority setting of corrective maintenance, and defect elimination are embedded in each of the above elements.

B2. **Maintenance**
Primary focus

The primary focus must be on safety-critical plant and equipment. These are plant and equipment that are relied upon to ensure safe containment of hazardous chemicals and stored energy, and for continued safe operation. This will typically include those items in a plant's proactive maintenance program, such as:
- Pressure vessels
- Piping systems
- Relief and vent devices
- Instruments
- Control systems
- Interlocks and emergency shutdown systems
- Mitigation systems
- Emergency response equipment
- Rotating machinery.

Proactive maintenance can be depicted as in Fig. B1.

There is a move toward total productive maintenance to improve integrity and production by means of proper care for equipment and production processes by empowering employees. The paragraph on *basic equipment care* discusses this further.

FIGURE B1

Proactive maintenance.

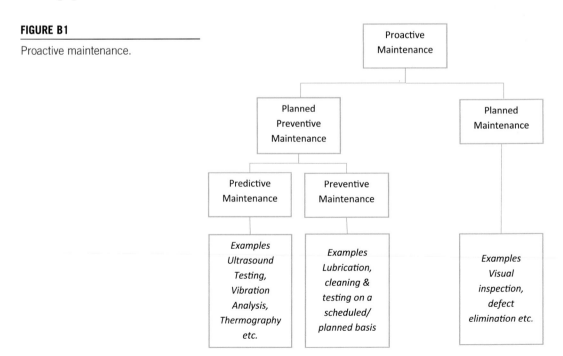

Standards

Some useful standards are:
- **SAE STD JA1011** Evaluation criteria for RCM processes
- **UK HSE RR 509** Plant Aging—Management of Equipment Containing Hazardous Fluids or Pressure
- **ISO 14224** Petroleum, Petro Chemical, and Natural Gas Industries—Collection and Exchange of Reliability and Maintenance Data for Equipment.

Maintenance principles

The primary maintenance principles are as follows:
1. Defect elimination
2. Optimum reliability and integrity
3. Optimum work volume
4. Maximum efficiency of execution.

Application of the primary maintenance principles:

Defect elimination reduces:
1) the **number of breakdowns** (emergency/breakdown maintenance) thereby increasing reliability and integrity,
2) **reactive work** (corrective maintenance) thereby optimizing work volume, and
3) **high priority work** thereby increasing efficiency.

Reliability and integrity improvement also enhances optimization of work volume, which in turn has the knock-on effect of improving efficiency of execution.

Active **defect elimination** is thus a key element of RCM, CBM, SIF, and RBI. Early detection includes visual inspection, nondestructive testing, etc., so that small issues can be tackled promptly before they can develop into something more serious.

A statistical approach to maintenance management, by means of effective data collection/study of industrial equipment behavior during its life span (for example, the number of breakdowns within a specified time window), needs to be applied. Computerized maintenance management system (CMMS) databases have the data and tools, such as Meridium, to analyze the data. Box B1 outlines active defect elimination utilizing a bad actor process.

Areas of effective maintenance

Areas of effective maintenance can be grouped:
- Maintenance management
- Turnaround management
- Projects

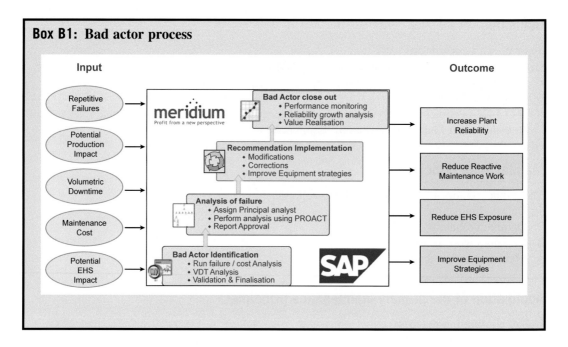

- Other
  a. Risk-based decision-making (see Section B3)
  b. Performance management (see Section B4)
  c. Initiative management (see Section B5).

Equipment criticality analysis

Criticality of equipment is directly related to the effect on production throughput.

Criticality is generally classified as A, B, and C.

A: Primary process stream equipment shutdown causes production loss.

Example:

If there is only one pump in the primary process stream and it fails, the process stops and production is lost. This pump would be regarded as a "criticality A" item of equipment.

B: Primary process stream equipment with parallel full capacity equipment.

Example:

Two crude feed pumps are installed in parallel, each capable of handling 100% of the refinery design capacity. One is electrically powered while the other is steam turbine driven. When power is lost to the plant, the steam-driven pump automatically cuts in. These pumps are regarded as "criticality B" items of equipment.

C: Equipment that could cause production loss.

Example:

Three air compressors supply the plant air system. One hundred percent of the air requirements are covered by two compressors. Failure of more than one air compressor would shut part of the plant down. These compressors are regarded as "criticality C" items of equipment.

Spares for criticality A equipment should always be retained in stores. These are referred to as insurance spares and would include items such as spare compressor and turbine rotors for criticality A equipment. Thus, maintenance, repair, and overhaul inventory management is critical to optimize stock but without allowing stock investment to get out of control. Spare part management techniques, such as LIFO (Last In First Out), FIFO (First In First Out), criticality analysis, etc., are used to prevent deterioration of stock and to ensure the correct spares and quantities are retained. Criticality analysis by matrix (risk vs. consequence of failure) is used for highly critical equipment, and analysis using criticality versus frequency of failure is used for spare parts.

Examples of equipment criticality are discussed in Box B2.

Planned maintenance intervals

Planned maintenance (PM) intervals are determined in the following order:
1. Original equipment manufacturers' (OEM) recommendations
2. Owner experience of this equipment based on RCM and RBI
3. Staff experience with this type of equipment.

Box B2: Equipment criticality

Compressors

Compressed air is critical for the smooth running of instruments that keep a refinery operating efficiently.

Two air compressors, each with 100% capacity, supplied instrument air for a certain refinery. One of these broke down and could not be repaired immediately since there were no spares in stock. As misfortune would have it, the other compressor also failed, causing the entire refinery to shut down. Production stopped completely for a number of hours, before a portable compressor was hired to keep the refinery online until the damaged compressors could be repaired.

Comment

At the time critical spares were not kept in stock since the compressors were rated as criticality C. However, since the lost production was significant, spares were subsequently kept in stock.

Another solution could have been to have a service agreement with the compressor supplier, where spares are required to be supplied within hours, and so mitigate such difficulties.

Relief valves

A gas plant experienced an increase in the backlog of repairs to relief valves. (Relief valves and their components are criticality A items.) An investigation was initiated to determine the reasons for the backlog, and it was found that there was an unacceptably long delivery time for these spares.

Outcome

Service agreements were established with the agents of all relief valve suppliers, whereby they guaranteed supply of spares, on demand, within a number of hours, and, as a result, the backlog was cleared.

General comment

The cost of such a service agreement (on a yearly basis, for example) versus cost of breakdown (i.e., production loss) needs to be evaluated to ascertain if it is cost effective. Depending on factors such as equipment criticality and careful cost analysis, the best compromise (economically speaking) should be adopted. It may turn out that having spare parts available on-site within 24 h by the supplier might be unacceptably expensive, whereas keeping stock in company stores is more viable.

Standards such as described in Box B4 also assist in determining PM intervals.

RCM and CBM processes

Failure analysis (root cause failure analysis or fault tree analysis) must be carried out on equipment that has failed in order to prevent repetitive failure. This process contributes toward **eliminating defects**.

Vibration analysis and monitoring of rotating equipment are key activities. Major equipment on primary systems requires online vibration measurement.

Lubricating oil analysis for key machines should be undertaken on a regular basis.

Example

Contaminants in the lubricating oil could predict items such as bearing failure.

Critical success factors (CSFs) for determining run time of critical rotating equipment:
■ Type categories (pumps, turbines, etc.)
■ Vibration analysis
■ Lubrication oil analysis
■ Failure analysis
■ Categorization of mean time between failures.

Electrical

It is necessary to comply with various International Electrotechnical Commission (IEC) standards.

CSFs for determining the electrical inspection intervals for equipment requiring process shutdown are:
■ List of equipment requiring a process shutdown
■ Analysis of required equipment shutdown intervals.

Key elements of good practice are:
• An extensive and effective lockout, tagout system controlled by the permit system
• Routine thermography of all transformers and switch-gear
• Electrical earth bonding circuits tested annually and postmaintenance break-in
• Maintenance of steam and gas turbine generator units in accordance with the OEM recommendations.

Instrument

IEC 61511 covers the design and management requirements for safety instrument systems (SISs) from cradle to grave. Its scope includes: initial concept, design, implementation, operation, and maintenance through to decommissioning.

IEC 61508 is a generic functional safety standard, providing the framework and core requirements for sector-specific standards. Three sector-specific standards have been released using the IEC 61508 framework: IEC 61511 (process), IEC 61513 (nuclear), and IEC 62061 (manufacturing/machineries).

IEC 61511 provides good engineering practices for the application of safety-instrumented systems in the process sector.

In line with **IEC 61511 and IEC 61508** standards on SISs, the primary focus is:
* Emergency shutdown (ESD)
* Fire and gas (F&G).

Safety instrument function is required to be carried out at regular intervals as determined by these standards. SIF involves determining safety integrity levels for instrument loops, which in turn predicts required testing intervals.

Critical success factors for determining periods between inspection of instrument systems are:
■ List of equipment requiring process shutdown
■ Analysis of required equipment shutdown intervals.

CSFs for determining periods between inspection for F&G and ESD systems are:
■ F&G and ESD systems requiring process shutdown
■ Analysis of required equipment shutdown intervals.

Basic equipment care

Analogy: If you own a car and you personally drive it and check the oil, water, and tire pressure on a regular basis, it is likely that you will spot any maintenance issues that arise even before your vehicle's mechanic is aware of them.

Similarly, operators are more familiar with the performance of the specific equipment that they operate than maintenance staff are, as they are there listening, seeing, and feeling on a daily basis. They also understand the criticality of each item of equipment. They could thus carry out minor work on the equipment such as attending to small leaks, general cleaning, clearing of blockages, and other duties so as to release maintenance staff to do more skilled work. However, it is important to note that the tasks allocated to operators must be clearly defined, interruptible, and of short duration, and operators must be properly trained to perform these tasks. Ideally a feeling of ownership needs to be inculcated into the way operators manage and care for their equipment.

The trend in some critical plant processes is to have maintenance staff on shift as a front line to provide a faster response when maintenance expertise is needed.

The outcome is that plant reliability improves as problems are identified and attended to at an early stage, before they grow to be much bigger complications.

B3. Inspection

Basis of risk based inspection

API RP 580 is the basis for establishing RBI in the hydrocarbon industry.

Scope of RBI

RBI relates to the whole pressure envelope and is grouped as follows:
1. Pressure vessels—all pressure-containing components
2. Boilers and heaters—pressurized components
3. Heat exchangers—shells, floating heads, channels, and bundles
4. Rotating equipment—pressure-containing components
5. Process piping—pipe and piping components
6. Storage tanks—atmospheric and pressurized
7. Pressure-relief devices.

Advantage of RBI

The primary advantage of RBI is that uninspectable risks are reduced. Fig. B2 shows the effect of using RBI.

Codes

RBI relies on the application of a number of codes. The US API and ASME codes for pressure containment of vessels, pipelines, and tanks are depicted in Fig. B3.

Processes
Remaining life calculations

Regular thickness measurement of all primary pressure systems, and the use of risk-based prediction software tools, can predict the extent of the remaining life of that system or item of equipment.

FIGURE B2

Advantage of riskbased inspection.

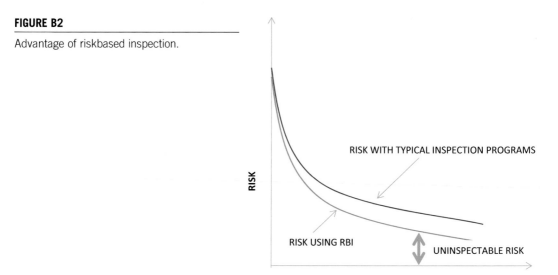

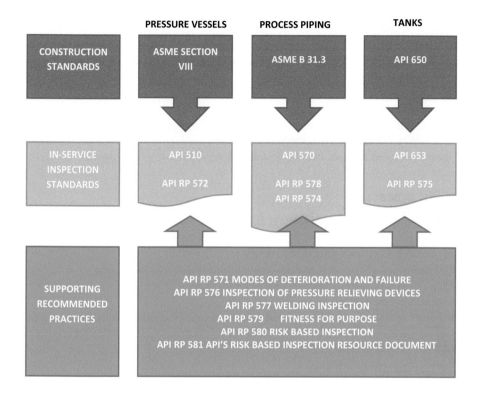

FIGURE B3

API codes involved with riskbased inspection.

Example

Remaining life calculations that take place during a turnaround can determine whether a new pressure vessel or column should be ordered for the next turnaround.

Nondestructive testing

Nondestructive testing covers all tools and methods used to assess the condition of the pressure systems without damage to these systems. These include magnetic particle testing, liquid penetrant testing, radiography (X-ray), ultrasonic thickness testing, thermography, electromagnetic testing, acoustic emission testing, guided wave testing, etc.

Destructive testing

Destructive testing (DT) results in irreversible damage to the tested item. It includes the testing of corroded/eroded items removed from the pressure systems when replacement takes place. An analysis may initiate a change of material for a particular service. Examples of DT techniques include stress tests, hardness tests (Charpy impact test), etc.

Corrosion management

Corrosion management includes the injection of corrosion inhibitors, impressed current, sacrificial anodes, and use of corrosion coupons to monitor and minimize corrosion of pressure systems and structures.

Example

Inadequate cleaning and passivation at start-up, and insufficient injection of corrosion inhibitors during operation, resulted in premature corrosion of a cooling water system in a polyester polymer plant. As a result, certain heat exchanger bundles required early replacement.

Output of RBI processes

The output of the inspection planning process, conducted according to API 580 RBI guidelines, should be an inspection plan for each item of equipment that will be analyzed, which includes:
1. Type of inspection methods that should be used
2. Extent of inspection (% of total area to be examined or specific locations)
3. Inspection interval or next inspection date (timing)
4. Other risk mitigation activities
5. Residual level of risk after inspection and other mitigation actions have been implemented.

An RBI application is discussed in Box B3.

Box B4 outlines **BP RP 32-3**, which became the de facto unofficial standard for many oil and gas companies and one of the foundations for later RBI standards.

Critical success factors (CSFs) for determining periods between inspection of static equipment:
■ List of equipment requiring process shutdown
■ Remaining life calculation
■ Recommended inspection intervals.

Box B3: Inspection—RBI application

An asset technical integrity audit revealed an excellent riskbased inspection management system (RBIMS) for all pressure containment within the process areas of a particular gas plant. However, none of the pipelines leading across the country to the terminal and downstream processing units had a system for assessing risk and inspection frequency.

To address this oversight, spreadsheet listings of cross-country pipelines and respective inspection intervals were improved with the intent to adopt a suitable risk modeling tool for each pipeline.

Comment

It stands to reason that risk assessment of all pressure-containing plant and equipment within the pressure envelope must be addressed. However, RBI, as defined by API 580, does not cover cross-country pipelines since these are covered by various modeling tools and techniques.[3]

> **Box B4: BP RP 32-3 Inspection and testing of in-service civil and mechanical plant—management principles october 1998**
>
> **Background**
> Industry is continually grappling with the pros and cons of extending run length between turnarounds using riskbased inspection. This standard sets intervals for plant inspection and testing based on experience.
>
> **Scope**
> Applicable to oil refineries, petrochemical and chemical plants, onshore and offshore production facilities, transmission pipelines, and storage and distribution facilities.
>
> **Value**
> Ensuring a high standard of plant integrity based on economical safe practice.

CSFs for determining periods between inspection of mechanical safety devices:
- List of devices requiring process shutdown
- Recommended inspection intervals.

B4. Software
Computerized maintenance management

CMMSs are now well established in the process industry, either as part of the enterprise resource planning (ERP) system, such as SAP or Oracle, or they have strong links to the company ERP system. Maintenance modules are linked to stores/purchasing and accounting modules so that the planning and scheduling of maintenance activities are streamlined. IBM Maximo and SAP EAM software[1] are the most common CMMSs in the oil, gas, and process industries.

The following are some advantages achieved with CMMS:
- Human resource restraints with regard to maintenance can be identified and leveled
- Material and tools availability can be identified from the materials module
- Financial (budget) restraints can be identified
- Work orders are able to be categorized according to emergency, corrective, and planned work
- Prioritization can be arranged based on equipment criticality
- Standard planned maintenance activities and intervals are embedded
- Multidiscipline job preparations are completed using detailed task breakdowns
- Efficiency is improved by monitoring "hands-on tool" time and waiting time for permit issue
- Monitoring of maintenance key performance indicators (KPIs) are efficiently enabled.

Box B5 outlines the evolution of CMMSs.

Master equipment list

The master equipment list for the plant resides in the CMMS and is accessible through the ERM system to the materials management system, inspection management system, and critical path models. Activity-based costing can also make use of this list to determine which items of equipment contribute the most to the profit of the company.

Box B5: Computerized maintenance management system evolution

In this day and age of computer technology, it is difficult to believe that in the 1970s at ICI, there was a Kardex system with each item of plant and equipment itemized on separate cards. Additionally, each card recorded the history of each and every item of equipment and this was either handwritten or typed.

When I joined Chevron in the 1980s, we had maintenance and planning system (MPAC) on the mainframe (BP had Maximo). When SAP enterprise management system came along with the maintenance and materials modules linked, the maintenance staff were reluctant to change from MPAC as they believed it had more to offer than the SAP maintenance module, which was deemed to be inferior and more generic. Nevertheless, the staff were eventually coerced into using SAP as it was considered to be for the greater good of the company.

SAP and Oracle (the leaders in ERP) consequently became the common ERP systems in the oil and gas industry. Oracle later bought PrimaVera Critical Path software (see Box C1). Maximo was bought by IBM. IBM Maximo and SAP EAM are now the leaders in CMMS.

Riskbased inspection

RBI software is essential for RBI. Box B6 outlines the application of Capstone[2] RBI software at Yara, the biggest fertilizer manufacturer in the world. Capstone has also been applied in various Chevron refineries.

Box B6: Capstone application press release[4] extract

Yara ready to roll out Lloyd's Register Capstone inspection package

Yara International ASA chose a solution from Lloyd's Register Capstone, Inc. to implement a riskbased inspection system at facilities throughout their company. In recent years Yara had tested different tools for RBI and had even developed and implemented its own RBI tool in Le Havre. However, they reached the conclusion that Reliability Based Mechanical Integrity (RBMI) from Lloyd's Register Capstone offered the best solution.

RBI is a departure from the traditional system, in which site and equipment inspections are imposed by outside authorities and are based on local regulations that often vary between countries. "Authorities tell you what, when, and how to carry out inspections, but without detailed knowledge of an installation. RBI is an inspection philosophy that is knowledge-rather than regulation-based. In the inspection world this is a big step forward."

Now Yara has opted for a forward-looking approach that sets an industry example. The RBI approach optimizes inspection intervals based on risk ranking and inspection methodologies that look at degradation factors like corrosion and cracking. The RBI approach on static equipment means increased safety and reliability, as well as reduced maintenance costs thanks to preventive, instead of corrective, actions.

"Now this process is a team effort from production, maintenance, and inspection—bringing the knowledge of all three together. In isolation you could have an upset in operations, which might cause trouble later, escape detection. With this increase in combined knowledge you have a thorough overview of what has happened in a plant."

RBMI has a certified link with SAP and is a complete tool, offering inspection plans and planning, a database for data collection and risks ranking, and an inspection data management system.

Asset performance management

Asset performance management (APM) software is becoming a necessity in the process industry.

Section 1.4.3 discusses the requirement for this, and Box 1.3 clearly shows the impact on other KPIs.

APM software should cover all assets in alignment with recognized and generally accepted good engineering practices (RAGAGEP) as per the following standards or similar:
- Reliability centered maintenance: SAE standard JA1011, "Evaluation Criteria for Reliability Centered Maintenance Processes"
- Safety instrument systems: ISA 84/IEC 61511 and IEC 61508
- Riskbased inspection: API 580
- Hazard analysis and operability studies (HAZOP/HAZAN): IEC 61882 "Hazards and Operability Studies Application Guide."

Focus areas
Strategy management
- Asset strategy manager
- Asset strategy optimization.

Strategy execution
- Calibration management in line with ISO 9000 and API Manual of Petroleum Measurement Standards
- Inspection management
- Operator rounds
- Thickness measurement location management including monitoring and thickness calculation according to ASME and API industry standards for piping, tanks, and pressure vessels.

Strategy evaluation
- Generation management
- Metrics and scorecards
- Production loss accounting
- Reliability analytics
- Root cause analysis
- Vibration analysis
- Asset criticality analysis.

Meridium/Predix is the most popular software covering these focus areas. Installations include Saudi Chemical, Bapco, Chevron, Qatargas, Rasgas, SABIC, Qatar Chemical, Saudi Aramco, and Exxon-Mobil. This is outlined in Box B7.

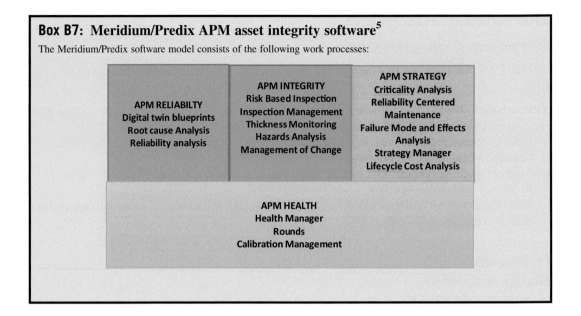

Box B7: Meridium/Predix APM asset integrity software[5]

The Meridium/Predix software model consists of the following work processes:

APM RELIABILTY
Digital twin blueprints
Root cause Analysis
Reliability analysis

APM INTEGRITY
Risk Based Inspection
Inspection Management
Thickness Monitoring
Hazards Analysis
Management of Change

APM STRATEGY
Criticality Analysis
Reliability Centered Maintenance
Failure Mode and Effects Analysis
Strategy Manager
Lifecycle Cost Analysis

APM HEALTH
Health Manager
Rounds
Calibration Management

Benefits of applying asset performance management software

Benefits of applying APM software include:
a) Improved capacity, utilization, and accuracy of forecasts
b) Balanced assessment focusing on the most critical and high-risk assets, using equipment risk profiles, production goals, and capital (CAPEX) and operating (OPEX) expenses (including maintenance and inventory costs)
c) Operational visibility and analysis that proactively reduces asset failures, controls costs, and increases production availability
d) Increased up-time and cost-efficient asset utilization while supporting repeatable best practices
e) Ensured compliance and safety by providing a comprehensive method to assess risks, reduce asset failures, increase process safety, control costs, and increase production availability.

B5. Organization

To maximize asset performance, cross-functional teams are required to take full advantage of the latest maintenance, inspection, and asset integrity software. An example of the integration is shown in Box B8.

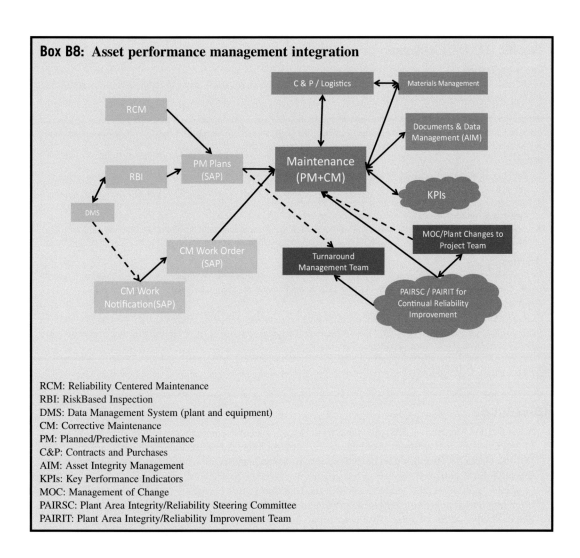

Box B8: **Asset performance management integration**

RCM: Reliability Centered Maintenance
RBI: RiskBased Inspection
DMS: Data Management System (plant and equipment)
CM: Corrective Maintenance
PM: Planned/Predictive Maintenance
C&P: Contracts and Purchases
AIM: Asset Integrity Management
KPIs: Key Performance Indicators
MOC: Management of Change
PAIRSC: Plant Area Integrity/Reliability Steering Committee
PAIRIT: Plant Area Integrity/Reliability Improvement Team

B6. Awards

The HART Plant of the Year is a unique award in the process automation industry, because it is the only public award presented to end user companies that recognizes ingenuity in the application of HART communication technology. The award showcases those end user companies and their suppliers who have demonstrated creativity in using the full capabilities of HART communication technology.

The background to this award is outlined in Box B9.

Box B9: HART/Fieldcomm promotion[6] extract

The organization

FieldComm Group began operations on January 1, 2015, by combining all assets of the former Fieldbus Foundation and HART Communication Foundation.

FieldComm Group is a global standards-based nonprofit member organization consisting of leading process end users, manufacturers, universities, and research organizations that work together to direct the development, incorporation, and implementation of communication technologies for the process industries.

Mission

1. Develop, manage, and promote global standards for integrating digital devices to on-site, mobile, and cloud-based systems
2. Provide services for standards conformance and implementation of process automation devices and systems that enable and improve reliability and multivendor interoperability
3. Lead the development of a unified information model of process automation field devices while building upon industry investment in the HART, FOUNDATION Fieldbus, and FDI standards

Overview

Deployed in nearly every industrial facility in the world, our technologies have provided a connected framework using intelligent field devices to reduce waste, improve safety, and increase operational efficiency for over 25 years.

Plant of the Year award

FieldComm Group's Plant of the Year is presented annually to an *end user company* for exceptional and innovative use of its technologies (FOUNDATION Fieldbus, HART Communication incl. wireless & IP, or FDI) in real-time applications to improve operations, lower costs, and increase availability.

References

1. Adair B. IBM Maximo vs. SAP. Available from: https://selecthub.com/cmms/eam/ibm-maximo-vs-sap/.
2. Lloyds Capstone RBI software. Available from: http://rbi.lrenergy.org/capstone-rbmi-software/.
3. Dey PK, Gupta SS. Risk-based model aids selection of pipeline inspection, maintenance strategies. *Oil & Gas Journal* 9 July, 2001.
4. Capstone Application 2007 (M2 PressWIRE via Thomson Dialog NewsEdge) RDATE:14122007. Available from: http://www.transformingnetworkinfrastructure.com/news/2007/12/17/3168231.htm.
5. BHGE Predix APM modules. Available from: http://solutions.geoilandgas.com/c/apm-asset-performanc?x=cFFgKt&utm_source=website&_ga=2.171080352.1697447539.1536988278-1179021930.1536988278.
6. Field Comm Group. Available from: https://fieldcommgroup.org/.

Further reading

1. Oliver R. *Turnarounds, an integral component of asset performance management*. World Refining; 2003.
2. Operations Integrity Management System (OIMS). ExxonMobil. Available from: https://corporate.exxonmobil.com/search?search=operations%20integrity%20management%20system.
3. GE/Meridium web site. Available from: www.ge.com/digital/asset-performance-management.
4. Enterprise APM v4 software on-stage demo. Available from: https://www.meridium.com/type/demo.
5. Peters R. *Reliable maintenance planning, estimating, and scheduling*. 1st ed. Gulf Professional Publishing; 2014.

6. Maximize return on capital investment with predictive maintenance. Hydrocarbon Processing. Available from: http://www.hydrocarbonprocessing.com/magazine/2018/may-2018/columns/digital-maximize-return-on-capital-investment-with-predictive-maintenance.

7. Golightly R. *Seeing into the future with prescriptive analytics: a new vision for asset performance management*. Aspen Technology. Available from: www.aspentech.com.

8. BHGE. Available from: http://solutions.geoilandgas.com/c/ge-digital-apm ?.

9. Frost and Sullivan White Paper the Future of Inspections: Technologies Converge to Transform Asset Integrity Management; 2018.

10. Hey RB. *Performance management for the oil, gas and process industries*. Elsevier; 2017.. In: *https://www.elsevier.com/books/performance-management-for-the-oil-gas-and-process-industries/hey/978-0-12-810446-0*.

Appendix C: Critical path method software

C.0 Introduction

The critical path method (CPM) has been used to manage complex projects since the 1960s. All project activities are sequenced in a model with single start and single finish dates. The method determines the shortest time to complete the project and is ideally suited to managing turnarounds.

Optimum downtime is determined by modeling all activities and determining and **minimizing the longest path of activities** with **optimum use of resources** required for the turnaround.

The model has to be built and maintained by an experienced planning engineer who has a thorough knowledge of the requisite software. A senior turnaround planning engineer should be the "second in-charge" of the turnaround. A capable senior turnaround planning engineer is "worth his weight in gold."

It is vital that the turnaround project manager and senior management understand the basics of critical path modeling as applied to turnarounds to ensure a successful end result. The key elements of the model are described in this appendix.

The software model depends on the complexity of the turnaround.

> *Example*
>
> *A certain nuclear power station has, in the past, used Microsoft Project for its turnarounds, whereas a number of refineries use either Primavera or Open Plan. A power station turnaround is inherently much simpler than a refinery or petrochemical plant.*

An alternative development to CPM, called program evaluation and review technique (PERT) uses a statistical technique to determine the activity duration. This was found to be very cumbersome since three estimates are required for an activity duration. As a result, CPM became the preferred method in new computer systems. However, PERT has subsequently been integrated with CPM in some software packages as an option for determining an activity duration (see Section C9). Both methods previously used the arrow diagram method (ADM), but later computer systems use the precedence diagram method (PDM) (see Section C1.1.1).

Fig. C1 shows a simple precedence diagram.

The evolution of CPM software is depicted in Box C1.

Box C2 discusses CPM software selection.

Box C3 describes the application of Primavera to refinery turnarounds.

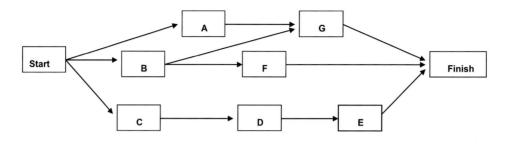

Precedent diagram method (PDM).

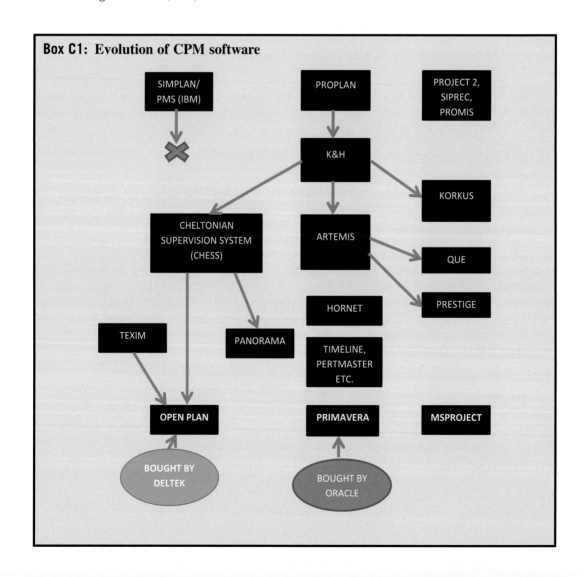

Box C2: CPM software selection

In the 1990s, my company selected Smartsuite for word processing, spreadsheets, database, and presentation. However, Microsoft muscled in and became a virtual world monopoly with Word, Excel, Access, and PowerPoint.

At the time, the company chose to migrate from Artemis (the world leader then in critical path modeling) to Open Plan, a PC-based system operating on the intranet. Artemis was mainframe based and was incurring high operating and IT support costs. By then PrimaVera and Open Plan had each cornered one-third of the market. These had a once-off cost of purchase requiring minimal IT support.

Downloading and editing of the models for a major refining complex from Artemis to Open Plan took a full year. These models are recycled regularly every four plus years, and upgraded and refined each time. Therefore, the safeguarding of these models in software that is not going to change or disappear in time is a challenge.

The selection of the most appropriate CPM software for a large complex is primarily dependent on the market share that the software commands, and the probability that it will still be prevalent in 10 or so years' time. Two software packages (Open Plan and Primavera) have now prevailed for more than 20 years. PrimaVera was acquired by the biggest ERP software company, Oracle, in 2008 and is currently the most popular high-level CPM software available for project management today.

Box C3: PrimaVera promotion

Seven refineries, one solution[8]

Marathon Petroleum's decision to standardize on one solution for multiple refineries is providing the integrated energy company with more efficient maintenance and turnaround and better use of its resources.

Marathon Petroleum Company (MPC) is the fifth largest refiner in the United States, with refining capacity of 974,000 barrels per day.

> *Going with one system makes our maintenance planning and operations more efficient …. In addition, our contractors use PrimaVera. It's standard in the industry, and it makes sense for us to adopt what the majority of people we work with use.*

Implementation of PrimaVera application for turnarounds

Agreement on structure of software model
1. Details for a basic structure for the system that would work at all refineries.
2. Make a list of the things that we all needed to agree on, such as resources, activity codes, calendars, and reports.

Implementation schedule
Week 1. Input into PrimaVera database from the history of each refinery's equipment: previous projects, maintenance and inspection records, etc.
Week 2. PrimaVera software training: 2 days basic and 3 days applied.
Week 3. Power user training: system administrators, planners, and others who use the system daily.
Week 4. Refinery personnel begin to use the software, with the implementation support team there to help.

PrimaVera is maintained by MPC, however, contractors can use the workstations at the company to assess requirements for their services, enter their requirements for resources, and plot their schedules.

Advantages. A planner can move from one refinery or turnaround project to another without having to learn a new planning system.

> *The biggest advantage in getting all seven refineries on the same planning software is the overall visibility MPC has gained into maintenance planning. This allows us to plan in a unified manner company wide, enabling us to contract for and deploy resources most efficiently.*

MPC "is the best in class in turnarounds."

Extract from Kreiling J. Seven refineries, one solution. Primavera Magazine.

C1. Theory of critical path method

The model entails a sequence of activities from a single start to a single end. Activities are both in series and in parallel. Activities are called tasks in Microsoft Project.

Each activity has predecessors and successors. Relationships could be finish-start, start-start, finish-finish, start–finish, start of hammock, and finish of hammock.

The critical path is calculated as the longest time of the sum of sequential activities from start to finish. First the forward pass calculates the total duration of each branch while the backward pass calculates the branch with zero slack. This is the critical path.

C1.1 Building a simple network diagram using the critical path method
C1.1.1 Network diagrams

Traditionally, two types of network diagrams have been produced: ADM and PDM.

ADM shows the activity in the arrow joining nodes. PDM shows the activity in a box with arrows showing relationships between boxes. This is currently the preferred method.

PDM is more specifically described as follows:
- A box or rectangle represents each activity
- Lines with arrows connect the boxes and represent the logical relationships between the activities
  - Predecessor—controls the start or finish of another activity
  - Successor—depends on the start or finish of another activity
- Start diagramming with either the first activity in the network and enter each successor, or start with the last activity in the network and enter each predecessor.

Note: There should only be **one start** and **one finish** for a network.

An example network is depicted in Fig. C2.

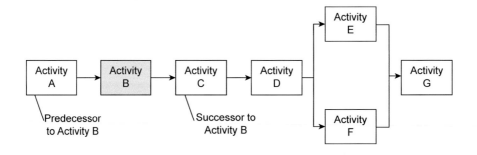

FIGURE C2

Simple precedence diagram.

C1.1.2 Identifying the critical path

The critical path method uses activity durations and relationships between activities to calculate schedule dates. This calculation is done in two passes through the activities in a project.

The critical path is the longest continuous path of activities through a project and determines the project completion date. A delay in one activity delays other activities and the project as a whole.

All projects should have a single start activity (start milestone) and a single finish activity (finish milestone).

a. Forward pass

The forward pass calculates an activity's early dates. Early dates are the earliest times an activity can start and finish once its predecessors have been completed. The calculation begins with the activity without predecessors (start activity milestone).

Calculation is: early start + duration − 1 = early finish.

This is depicted in Fig. C3.

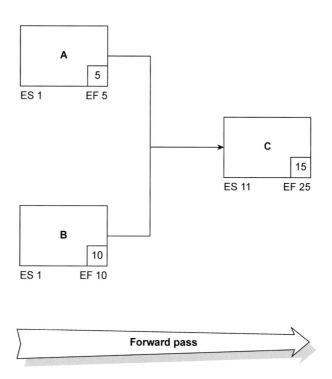

FIGURE C3

Forward pass.

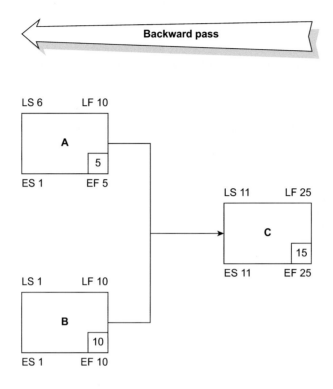

FIGURE C4
Backward pass.

b. Backward pass

The backward pass calculates an activity's late dates. Late dates are the latest times an activity can start and finish without delaying the end date of the project. The calculation begins with the activity without successors (finish activity milestone).

Calculation is: Late finish − duration +1 = late start.

This is depicted in Fig. C4.

C1.1.3 Total float

Total float is the amount of time an activity can slip from its early start without delaying the project. It is the difference between an activity's late dates and early dates. **Activities with zero total float are critical.**

Total float is depicted in Fig. C5.

Calculation is: late date − early date = total float.

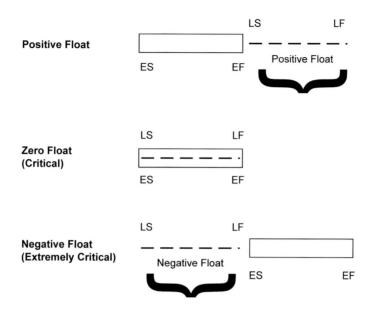

FIGURE C5

Total float.

Total float calculation is depicted as in Fig. C6.

A common project scenario is to set a required finish date for the project (a constraint), which is used only during the backward pass. The required finish date specifies when the project must be completed regardless of the network's duration and logic.

Calculation is: late finish − duration +1 = late start.

This is depicted in Fig. C7.

If the logic is not complete, a loop will occur. Loops indicate circular logic between two activities and the schedule calculation cannot proceed until all loops have been eliminated.

C1.2 Hammocks (open plan)

A hammock is an activity that does not influence the time analysis of a project but provides a means for reporting on start and finish of a group of related activities.

C2. The computer model

The model entails the following:
- Activities (see Section C3)
- CPM (see Section C1)

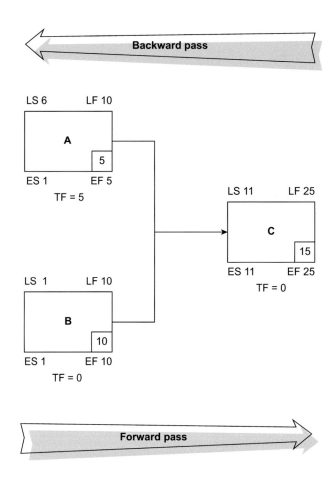

FIGURE C6

Calculation of total float.

- Resource scheduling/leveling (see Section C4)
- Earned value method (see Sections C6 and A4.4)
- Links to enterprise resource management systems—SAP, Oracle, etc. (see Section B4)
  - Maintenance Management Systems
  - Materials Management System
  - Inspection Management System
  - Financial (Budget) Management System
  - Clocking (Time) Management System
- Risk analysis (some products) (see Section C9).

The heart of the model is the activity or task to be done.

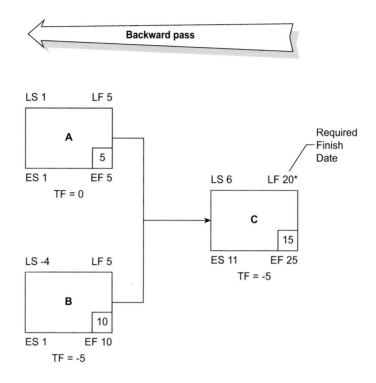

FIGURE C7

Calculation with finish restraint.

The main components of an activity are shown in Fig. C8.

Key components of the model

- Organizational breakdown structure (OBS)
    - Includes all line management
    - OBS chart, which is distributed at start of turnaround for posting on walls so everyone knows who is responsible for what
    - 48 h "look aheads" are easily printed for those responsible for action
    - Support staff that are considered to be resources (in terms of cost)
- Work breakdown structure (WBS)
    - All work must be under this hierarchy by area/discipline/system
- Cost breakdown structure (CBS)
    - Every activity must have a cost (except milestones)
- Contractor breakdown structure (may be part of OBS or separate)
- Calendars
- Resources.

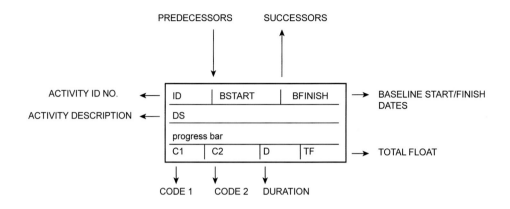

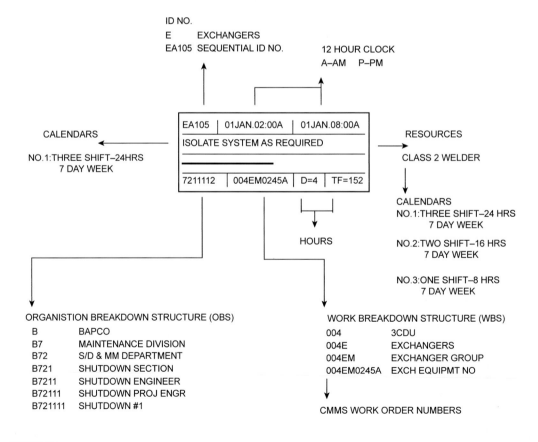

FIGURE C8

Main components of an activity.

Box C4: Organization breakdown structure example (Bapco)

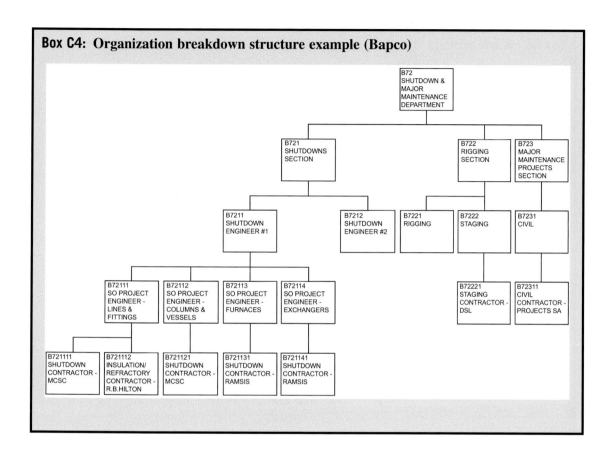

Resourcing must be integrated. Box C4 gives an example of an OBS with common rigging, staging, and civil resources.

Box C5 shows a WBS for a sulfur plant turnaround.

Each activity must be linked to a calendar. Ideally a calendar for maximum worker efficiency would be 10 h shifts with 2 h between shifts for handover meetings, progress updates, and work that does not allow people in the area, such as X-ray work and heavy lifts. An ideal week is a 6-day week with one rest day. This could be used to catch up if progress slips. However, critical and near critical activities may require work to be carried out 24/7 in, say, two 12 h shifts, and a 7-day week.

Examples of calendars are shown in Box C6.

Resources are attached to activities. Examples are shown in Box C7.

Box C5: Work breakdown structure example

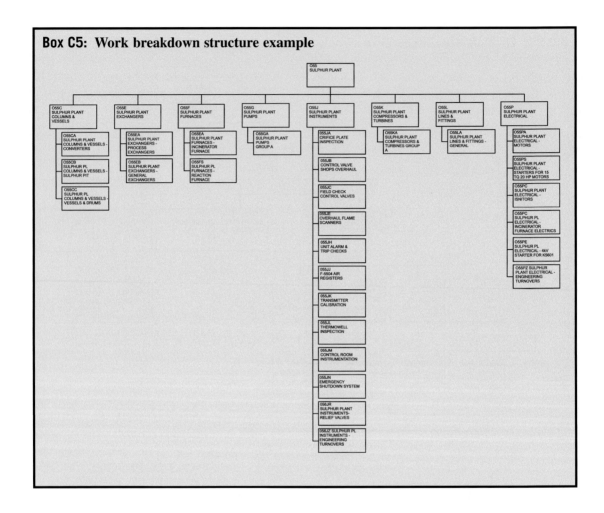

A resource may also need a calendar that could supersede the activity calendar.

Example

A crane activity is scheduled for the middle of the night using the activity calendar. However, crane lifts can only take place in daylight hours, so the resource calendar forces the crane activity to be carried out the next day.

Each activity could have other codes attached for ease of grouping and filtering when producing reports. Priority codes could also be applied for resource leveling (see Section C5). Box C8 lists other possible codes.

Activities are grouped into subprojects (fragnets in Primavera terms). These then make up a complete process unit shutdown network. This unit network is then integrated into a refining complex network. An example is shown in Box C9.

Box C6: Calendars

CODE	DESCRIPTION
7D24H	7 DAYS,24 HOUR DAY
7D24HL	7 DAYS,24 HOUR DAY WITH LUNCH
7D12HD	7 DAYS,12 HOUR DAY,DAY SHIFT
7D12HN	7 DAYS,12 HOUR DAY,NIGHT SHIFT
7D12HDL	7 DAYS,12 HOUR DAY WITH LUNCH
7D12HD6	7 DAYS,12 HOUR DAY SHIFT 6-6
7D12HN6	7 DAYS,12 HOUR NIGHT SHIFT 6-6
7D12HD8	7 DAYS,12 HOUR DAY 8-8
7D12HD8L	7 DAYS,12 HOUR DAY 8-8 WITH LUNCH
7DDL	7 DAYS,DA YLIGHT HOURS
7D11HN	7 DAYS,11 HOUR DAY,NIGHT SHIFT
7D11HD	7 DAYS,11 HOUR DAY,DAY SHIFT
7D2HB	7 DAYS,2 HOUR X RAYING
7D8HD	7 DAYS,8 HOUR DAY,DAY SHIFT
7D8HA	7 DAYS,8 HOUR DAY,AFTERNOON SHIFT
7D8HN	7 DAYS,8 HOUR DAY,NIGHT SHIFT
6D24H	6 DAYS,24 HOUR DAY
6D12HD	6 DAYS,12 HOUR DAY,DAY SHIFT
6D12HN	6 DAYS,12 HOUR DAY,NIGHT SHIFT
6D8HD	6 DAYS,8 HOUR DAY,DAY SHIFT
6D8HA	6 DAYS,8 HOUR DAY,AFTERNOON SHIFT
6D8HN	6 DAYS,8 HOUR DAY,NIGHT SHIFT
5D12HD8L	5 DAYS,12 HOUR DAY 8-8 WITH LUNCH
5D8HD	5 DAYS,8 HOUR DAY
5D8HDL	5 DAYS,8 HOUR DAY WITH LUNCH

The example shown in Box C9 depicts the following:
1. Oil out to oil in
2. Each unit scheduled separately
3. OBS ensures 48-h look ahead by area supervisor or common specialist service such as cranage
4. WBS can give progress on type of equipment
5. CBS can give progress on costs by area, contractor, or other
6. Resources leveled for whole model to check for overload of common resources such as cranes
7. Critical and near-critical activities reviewed "critically" with respect to time and resourcing to attempt to reduce critical path
8. Box notation shows one diagonal line through a box to show work in progress, a cross through a box for completed work and "nibbed" sides indicating a parent activity for a network diagram (subproject) below the box.

Box C7: Resources

RESOURCES		File No: CALRES.OPR	
		(example :FLM .BM)	
CONTRACTOR		*LABOUR*	
Code	Description	Code	Description
FLP	FLUOR PROJECTS	BM	BOILERMAKER
FLM	FLUOR MAINTENANCE	GF	GENERAL FITTER
SAF	SA FIVE	PF	PIPE FITTER
CBI	CHICAGO BRIDGE	FM	FIELD MACHINIST
BAB	BABCOCK	WE	WELDER
PIO	PIONEER	WX	CR WELDER
STE	STEINMULLER	HP	HELPER
ROT	ROTECH	LB	LABOURER
GLO	GLOBE	INSP	INSPECTOR
NAU	NAUTILUS	EN	ELECTRICIAN
SGB	SGB	EH	ELEC HELPER
CC	CONCOR	AL	ANALYSER TECH
CON	CONSANI	IN	INSTRUMENT TECH
LOH	LOHAR	IH	INSTR HELPER
CS	CONCORD	SA	SAFETY OFFICER
BW	B&W ELECTRICAL	HW	HOLEWATCHER
PRO	PROGRESS INSTR	FW	FIREWATCHER
STD	STANDARD TEMPLATE	SU	SUPERVISOR
		LH	LEAD HAND
		AC	AIRCON MECH
		TU	TURNER
DEPARTMENTS		*CREWS*	
Code	Description	Code	Description
RS	REL SERV	SVC	PSVCREW
EP	ENG & PROJS	HJC	HYDROJET CR EW
		STC	STAGING CREW
		BC	BURNER CREW
		LC	LAGGING CREW
		SRC	STRESS RELI EF CR
		DC	DUMPING CREW
		SBC	SAND BLAST CREW
		RFC	REFRAC/GUNITE CR
		RC	RIGGING CREW
		SC	STACK CREW
		SPC	STACK PAINTING CR
		PC	PAINTING CREW
		EQUIPMENT	
		Code	Description
		C10	10 TE CRANE
		C20	20 TE CRANE
		C40	40 TE CRANE
		CA	AMERICAN
		CMC	COMPRESSOR
		WMC	WELDING M/C
		SMC	SCREENING M/C
		SS	SUPER SUCKER
		TP	TEST PUMP

Box C8: Codes

C2 FIELD-SDOBS	File No: SDOBS.COD
SM	SHUTDOWN MANAGER
SM .AM	AREA MANAGER
SM.AM .XS	CALTEX SUPERVISOR
SM .AM .XS.CS	CONTRACTOR SUPERVISOR
C3 FIELD-WO No.	
123456789	EQUIPMT WO NO (10 ch.field)
C4 FIELD-PROD/SERVICE	
USH	HP STEAM
USM	MP STEAM
UFG	FLARE GAS
UBFO	BURNER FUEL OIL
UA	AIR
UC	CONDENSATE
C5 FIELD-QCP No.	
12 Character Field	
C6 FIELD-SCOPE CHANGE	
N prefix with sequential numbering	
C7 FIELD-MONITOR HAZARD	
A	Asbestos

C3. Creating network activities

Establishing an activity duration based on specific resources and calendars.

What are activities?

Activities are the fundamental work elements of a project. They are the smallest subdivision that directly concerns the turnaround project manager. (In Microsoft Project these are referred to as tasks.)

An activity:
- Is the most detailed work unit tracked in a project schedule
- Contains all information about the work to be performed.

Each activity requires:
1. Activity ID and description
2. Activity type
3. Duration
4. Dates and times

Box C9: Partial CPM network diagram of a refinery complex turnaround

Note: The detailed network diagram of the complete fluid catalytic cracking unit (FCCU) turnaround is under the **red** activity block. **Red** indicates that the FCCU is on the critical path for the whole refinery complex turnaround.

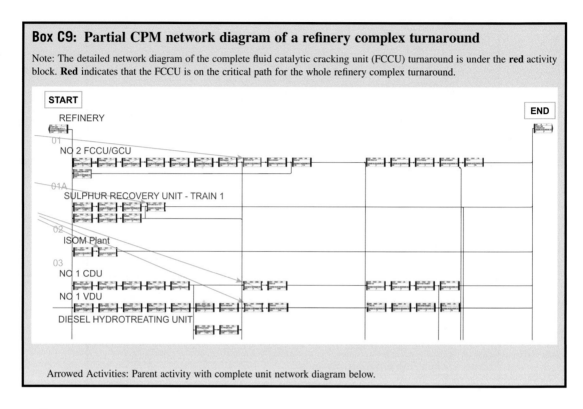

Arrowed Activities: Parent activity with complete unit network diagram below.

---

5. Calendar assignment
6. Duration and percent complete types
7. Work breakdown structure
8. Organization breakdown structure
9. Costs
10. Resource assignments
11. Constraints
12. Relationships
13. Activity codes.

1. Activity ID and description

Common practice for turnarounds is to use the type of work as the first descriptive of the ID and increase the number by 5 or 10.

For simple logic groups use two or three characters (e.g., activities for removal and replacement of an exchanger).

For groups of equipment use four characters, the first being in alphabetical order (e.g., a nest of exchangers).

Example type codes are shown in Fig. C9.

A   Pre-Turnaround
C   Columns and Vessels
D   Tanks
E   Heat Exchangers (including fin fans, cooler boxes and cooling towers)
F   Furnaces and Boilers
G   Pumps
H   Ejectors
J   Instruments
K   Compressors and Turbines
L   Lines and Fittings
P   Electrical
Q   Civils / Structural
R   Reactors
Z   Start up

FIGURE C9

Type of work codes.

The description should start with a **verb** to describe the activity.

Example

F005   **Start** *furnace repairs.*
F010   **Spade** *furnace tubes.*
F015   **Remove** *furnace man-ways.*

Avoid using I, O, and S for ID numbers as these letters are easily confused with 1, 0, and 5.

2. Activity type (PrimaVera)

The activity type controls how PrimaVera calculates an activity's duration/dates.

Types are:

Task dependent (default type in PrimaVera)

This type is typically used when the work needs to be accomplished in a given time frame, regardless of the availability of assigned resources. The activity's resources are scheduled to work according to the activity calendar. Duration is determined by the assigned resource calendar's work hours.

Resource dependent

This type is typically used when multiple resources assigned to the same activity can work independently. The activity's resources are scheduled according to the individual resource's calendar. Duration is determined by the availability of the resources assigned to work on the activity.

Level of effort

This type is typically used for ongoing tasks dependent on activities. Duration is determined by its predecessor/successor activities. Examples include meetings and project management tasks. Constraints cannot be assigned.

Start milestone

This type is typically used to mark the beginning of a phase or to communicate project deliverables. It is a zero duration activity and only has a start date. Constraints, expenses, work products, and documents can be assigned. A primary resource can be selected but roles cannot.

Finish milestone

This type is typically used to mark the end of a phase or to communicate project deliverables. It is a zero duration activity and only has a finish date. Constraints, expenses, work products, and documents can be assigned. A primary resource can be selected but roles cannot.

3. Duration

For turnaround execution, the duration is in **hours**. Other activities prior to, and after, the turnaround execution phase could have a duration in days.

4. Dates and times

Activity dates/times are:
- Early start
- Early finish
- Late start
- Late finish
- Scheduled start
- Scheduled finish
- Actual start
- Actual finish.

5. Calendar assignment

A file of calendars is created for selection of a suitable calendar for each activity. Some resources may also require an overriding calendar (see example in Section 5.3.1). Typical turnaround calendars are shown in Box C6.

Examples
- *7D24H can be used for curing of concrete.*
- *7DDL can be used for heavy crane lifts, which are required to be executed in daylight hours.*
- *7D2HB is suitable for X-ray work late at night when there is minimal activity.*

6. Duration and percent complete types

There are two alternatives, percent complete and remaining duration. Percent complete gives an accurate assessment of completion when the activity is, for example, "repair 10 valves." However, it gives an optimistic view when used for other activities and **remaining duration is then preferred**.

7. Work breakdown structure

For turnarounds, WBS is based on the process units and types of equipment to be worked on. The WBS is hierarchical. The general grouping for creating the WBS is:

Complex.

    Process unit.

        Equipment type/group.

           Equipment number.

At process unit level, which is generally a subproject, code levels are entered into the CPM software to create the WBS. Box C10 gives an example.

Typical equipment groups for a refinery or gas plant are:
- Columns and vessels
- Reactors
- Exchangers
- Furnaces
- Pumps
- Compressors and turbines
- Lines and fittings
- Electrical
- Instruments.

Box C10: WBS levels example

Level 1:	*3 Digits*	*Plant number*	*004*	*4 Crude distillation unit (CDU)*
Level 2:	*4 Digits*	*Equipment type*	*004E*	*4 CDU Exchangers*
Level 3:	*5 Digits*	*Equipment group*	*004EA*	*4 CDU Atmospheric overhead exchangers*
Level 4:	*10 Digits*	*Equipment number*	*004EA0221A*	*4 CDU Atmospheric overhead feed exchanger*

Box C11: Bapco turnaround OBS levels example

Level 1	1 Digit	Bapco	B	Company
Level 2	2 Digits	Division	B7	Maintenance division
Level 3	3 Digits	Department	B72	SD&MM department
Level 4	4 Digits	Section	B721	Shutdown section
Level 5	5 Digits	Group	B7211	SD engineer in charge of SD
Level 6	6 Digits	Subgroup	B72111	SD engineer/supervisor
Level 7	7 Digits	Contractor	B721111	MCSC contracting company

Box C12: Chevron Cape Town Refinery turnaround OBS levels example

C2 FIELD-OBS	File No: REFOBS.COD
1	REFINERY MANAGMT
1 .1	RELIABILITY SERVICES
1 .1 .1	CALTEX SUPERVISOR
1 .1 .1 .1	CONTRACTOR SUPERVISOR
C2 FIELD-SDOBS	File No: SDOBS.COD
SM	SHUTDOWN MANAGER
SM .AM	AREA MANAGER
SM .AM .XS	CALTEX SUPERVISOR
SM .AM .XS .CS	CONTRACTOR SUPERVISOR

---

8. Organization breakdown structure

OBS must include all lines of command and are thus hierarchical. Box C11 shows the standard structure used in Bahrain Petroleum Company (Bapco).

Box C12 shows the standard OBS structure for Chevron Cape Town Refinery as entered into the CPM model.

9. Costs

Cost reporting

Activity costs are displayed in three categories:
1. Labor
2. Nonlabor (equipment and materials)
3. Expense.

Progress reporting is displayed using the **earned value method**.

Planning costs

Costs are planned and managed at the activity level. Generally, there are two types:
1. Unit price, which is calculated based on resource assignments.
2. Lump sum costs that are manually entered into the model.

Unit price

The cost of a resource is based on the price/unit defined in the resource dictionary.

Budget cost = Budgeted units × Price per unit.

Expenses

Expenses are nonresource costs associated with a project. They are typically one-time expenditures for nonreusable items.

Examples include materials, travel, facilities, overheads, and training.

Monitoring and controlling costs

Unit costs are automatically calculated based on the actual number of hours entered against an activity. These are calculated from the actual start date of the activity and the reporting date for progress.

Earned value is determined when the percent complete and/or the remaining duration is entered.

10. *Resource assignments*

Resources are divided into two categories: labor and nonlabor.
 I. Labor (people) is time based, generally reused between activities and recorded in terms of price per unit
    Example: $10/h in an 8 h shift.

 II. Nonlabor (materials and equipment) are recorded in price per unit
    Example: $4 per square meter.

Typical resource codes are shown in Box C7.

11. *Constraints*

Constraints are imposed date restrictions used to reflect project requirements that cannot be built into the logic. Constraints are user imposed, and two constraints can be assigned to an activity.

After applying a constraint, the project must be rescheduled to recalculate new dates.

Benefits:
- Enables building a schedule that more accurately reflects the real-world aspects of the project
- Provides more control of the project
- Can be used to impose a restriction on the entire project or on an individual activity.

Commonly used constraints
- Must finish by
- Start on or after
- Finish on or before.

Other constraints
- Start on
- Start on or before
- Finish on
- Finish on or after
- As late as possible
- Mandatory start and finish.

Note: Constraints should be used sparingly as the model might behave bizarrely, especially if some imposed restraints have been forgotten.

12. Relationships

As mentioned earlier, PDM is used. Activities have both predecessors and successors. The predecessor controls the start or finish of another activity while the successor depends on the start or finish of another activity.

Relationship types
- Finish to start (FS)
- Start to start (SS)
- Finish to finish (FF)
- Start to finish (SF).

Relationships with lag

Lag specifies an offset or delay between an activity and its successor. It is expressed in the same duration units as the activities (hours in the case of turnarounds), is scheduled based on the calendar of the successor activity, can be added to any type of relationship, and can have a positive or negative value.

Relationships are:
- Finish to start with lag
- Start to start with lag.

> *Example*
>
> *Once concrete has been poured, a setting time is required before any other activity can take place on the concrete. Thus a lag for "setting time" can be imposed at the end of "pour concrete."*

13. Activity codes

Additional codes can be attached to the activity. These could include reference to documents such as work orders, quality control procedures (QCPs), drawings, etc., or be used for logging scope changes or hazardous operations. Reports can then be filtered based on these codes.

Examples are shown in Box C8.

Activity work packs

Activity work packs are used to build the CPM model and give details of what is required in the field.

Work is categorized before being deconstructed into a network of activities. Categories are:
1. Major tasks
2. Minor tasks
3. Bulk work.

The development of these work packages is discussed in Section 4.3.3.

C4. Resource scheduling and leveling
What is resource scheduling?

Resource scheduling is the process of calculating a project schedule based on the availability of resources required by the activities.

The purpose of resource scheduling is to calculate a project schedule that takes into account the limited availability of resources as well as the durations, logical relationships, and target dates of the activities.

The following topics are discussed:
1. Resource scheduling methods
2. The results of resource scheduling calculations
3. Resource scheduling rules.

Resource scheduling methods

Two basic methods are:
1. Time-limited
2. Resource-limited

1. Time-limited resource scheduling

Time-limited resource scheduling places a priority on maintaining the overall project completion date while attempting to minimize the extent to which any resource is overutilized. In time-limited resource scheduling, constraints on resource availabilities are not allowed to delay the overall completion date of the project. As a result, resources may be overloaded, if necessary, to prevent an activity from finishing after its late finish date.

Time-limited resource scheduling places all scheduled dates between the early and late dates of activities. Thus only activities not on the project critical path can be delayed by resource constraints.

In order to minimize overutilization of resources without interfering with the project completion date, resource leveling is applied within time-limited resource scheduling. In other words, the start and finish dates of an activity are moved within the free float of the activity to reduce the resource peaks to within a set limit, if possible.

2. Resource-limited resource scheduling

Resource-limited resource scheduling places a priority on preventing the overutilization of resources, even if that means putting the project behind schedule. In the absence of thresholds, or immediate activities (discussed later), resource-limited resource scheduling delays project completion, if necessary, to ensure that the resource requirements do not exceed availabilities.

The results of resource scheduling calculations

Resource scheduling calculates:
- Scheduled dates
- Earliest feasible date
- Scheduled duration
- Delaying resource
- Scheduled float.

Resource scheduling rules

The general rules of resource scheduling:
1. Activities are scheduled one at a time in series.
2. Activities before "time now" are not scheduled.
3. An activity is not scheduled before its early start date.
4. Activity durations are never shortened, but may be lengthened due to stretching, splitting, or reprofiling.
5. Resource scheduling cannot violate project logic.
6. Time-limited resource scheduling schedules within the late finish date of the project if there is no negative float. If negative float is present, the model schedules on early finish date of the affected activity.
7. Resource-limited resource scheduling does not consider an availability in excess of the defined resource availability except for immediate activities (which are scheduled on the earliest feasible date whether or not resources are available) or when thresholds are used.

Resource scheduling attributes for activities

Splitting, stretching, and reprofiling are three activity attributes that can be used to add flexibility to an activity during resource scheduling. Each of these techniques can be regarded as a way to improve the performance of a project by relaxing the rules of resource scheduling in a particular way.

It is also possible to assign an "immediate" attribute to an activity. The "immediate" attribute removes flexibility from resource scheduling calculations by forcing the model to schedule an activity on the earliest date possible.

C5. Activity prioritization[1]

The main objective at the start of a turnaround is to open up the equipment as soon as possible to determine how much emergent work may be required. (The inspection department could advise which equipment is most likely to have emergent work.) However, due to resource constraints, all the equipment cannot be opened at once, so priorities need to be set.

The critical path is usually quite straightforward and, on the model, many tasks are set sequentially and are therefore inflexible.

To maximize early activities, the model needs to be built with as many parallel activities and as few sequential ones as possible so that the scheduling tool can allocate tasks where required.

Allocating high priorities to equipment will allow the scheduling tools resource leveling capability to drive these first. Otherwise the software could prioritize reassembling of the first priority equipment before opening of the next piece of priority equipment for inspection.

In the example in Fig. C10, the isolation, opening, and inspection of the equipment is prioritized first by allocating numbers 1–3 (1 being the most critical), then the task for potential emergent work as number 4, followed by numbers 5–7 for the close-up and reassembly.

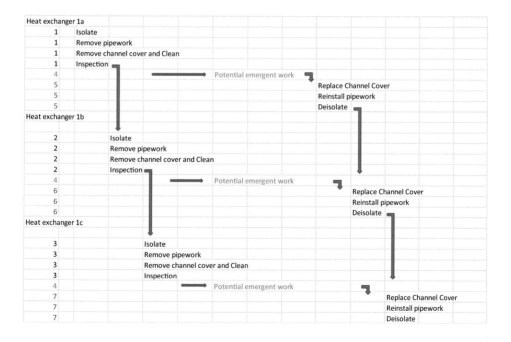

FIGURE C10

Prioritization explained.

This simple approach allows the scheduling tool to prioritize the work within the available resources and relationship links.

C6. Schedule optimization and risk assessment

Schedule optimization is the most crucial step prior to freezing the scope. It is vital that both the core team and main contractor management are totally involved and understand the background to the behavior of the model. All activities for the particular turnaround must be included in the model.

Schedule optimization includes:
- Analysis of the model, which must include a thorough evaluation of the critical path and near-critical path activities. The number of near-critical path activities can affect the probability of achieving the planned completion date.
- Probability analysis of **high-risk activities** and the influence on the completion date, which could be performed if the model is well structured.
- Activity prioritization (see Section C5).

The schedule optimization process is as follows:
- Run model
  - Time restraint
  - Resource restraint
- Review all
  - Critical and near-critical activities
  - Long-duration activities
  - High-risk (known unknowns) activities
    - *Example: column minor repack, partial repack, or full repack*
  - High priority activities
- Edit/rerun until acceptable
  - Options
    - Shift work
      Example: X-ray on night shift
    - Resource loading: contractor limits/location limits
    - Change work method/sequence
    - Parallel work
    - Contingency
    - Emergent work factor
- Freeze
  - Set baseline
  - Obtain final approval
  - Issue initial charts/plans/reports.

Some software packages allow the use of the PERT methodology for certain activities to assess the probability of achieving the targeted duration for the project (see Section C9 for examples).

C7. **Earned value**

Earned value determines progress of activity work relative to the baseline and how effective actual man-hours are relative to planned man-hours. Productivity is determined differently and is related to total actual activity work hours relative to the total clocked time.

Key variances are:
1. Schedule variance
   a) Where you are relative to where you should be at this time
2. Cost variance (CV)
   a) How much you have spent relative to how much you should have spent at this time
3. Estimate at completion
   a) Calculated from CV

Earned value is discussed in more detail in Appendix A4.4.

C8. **Model updates**

Once the scope is agreed, the model baseline is set. The baseline is the agreed schedule for the turnaround, and progress is then measured against this baseline.

Daily progress reporting for entry into the critical path model includes:
1. **Number of items completed** out of total planned number of items within an activity
2. **% completion** or **remaining hours.**

Actual hours are automatically calculated.

C9. **Risk management**

Risk management relates primarily to health, safety, environment, quality, duration, and cost. Duration and cost risk management may be available in, or linked to, some critical path software.[2,3] Probability analysis is carried out in some cases. Graphical displays include:
1) Project completion date probability
2) Project completion cost probability
3) Project duration probability.

Project completion date probability shown in Box C13.

Project completion cost probability is shown in Box C14.

Project duration probability is shown in Box C15.

Box C13: Project completion date probability chart example

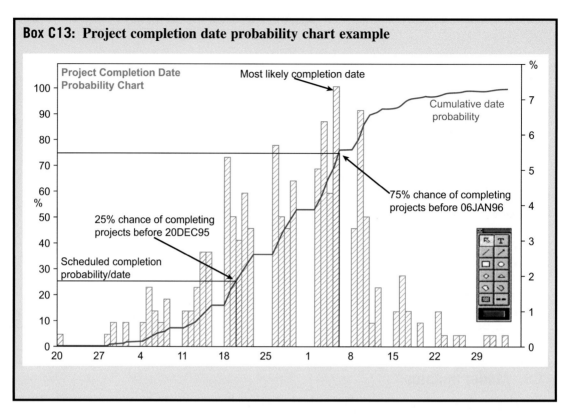

Box C14: Project completion cost probability example

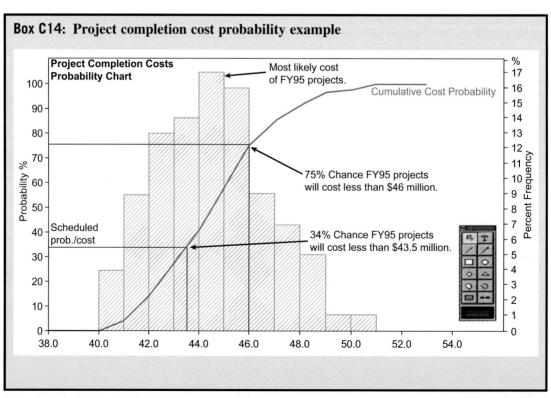

Box C15: Project duration probability example

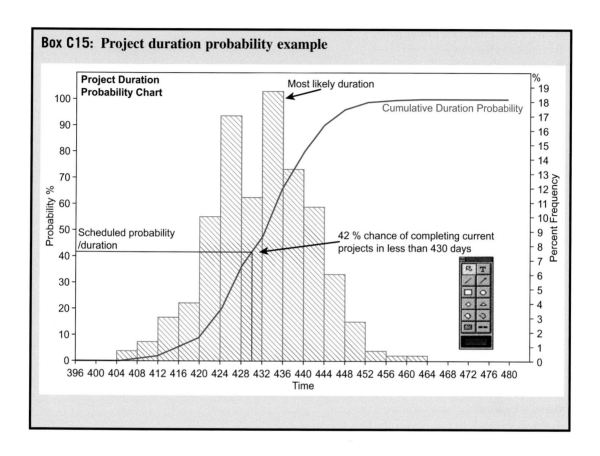

C10. Reports

Numerous reports can be produced and tailored to suit the relevant recipients. The primary views are:
1. Network
2. Bar chart (Gantt)
3. Spreadsheet (table)
4. Histogram
5. S-curve.

Items two to five are often combined. Bar charts sometimes show linkage between activities.

Reports include:

General viewing in the shutdown activity center:
1. Skeleton network
2. Summary Gantt charts by process units
3. 48 h look ahead for near-critical and critical activities for each unit

4. Resource histogram (overall)
5. Earned value curves
6. Productivity curves.

Top management:
1. Earned value curves—most valuable for seeing the big picture
2. Summary Gantt charts by process unit
3. 48 h look ahead for near critical and critical activities for each unit
4. Productivity curves
5. Exception reports: listing activities behind schedule, cost overruns, additional work and costs, quality excursions, and HSE incidents.

Area supervisors:
1. 48 h look ahead for all work for which they are responsible
2. Histograms for area resources
3. Construction packages including drawings, specifications, and QCPs

Box C16 shows a typical project schedule as a Gantt chart with its associated resource histogram.

Box C17 shows an example of a resource histogram for all labor for a process unit turnaround where progress is behind schedule.

C11. Critical path modeling benefits

Primary benefits:
1. Model/s archived and copied for next turnaround
2. Work scope and resourcing can be produced quickly for emergency shutdowns from archived model/s, and opportunity work can be slotted into the emergency work
3. Negates the need for building from scratch for each turnaround
4. Progressive improvement of model for each successive turnaround
5. Daily updates on progress and cost during the execution phase.

C12. Support software[4–7]

Software that links and supports the critical path software could be useful when preparing, planning, and/or reporting progress. Primary benefits relate to work flow and cost management. Box C18 gives an example.

C13. Summary

Complex turnarounds require the construction of an integrated CPM. Experienced planning engineers, who can use the software effectively, are required. Models/components must be archived for future turnarounds.

Box C16: Gantt chart and resource histogram example[9]

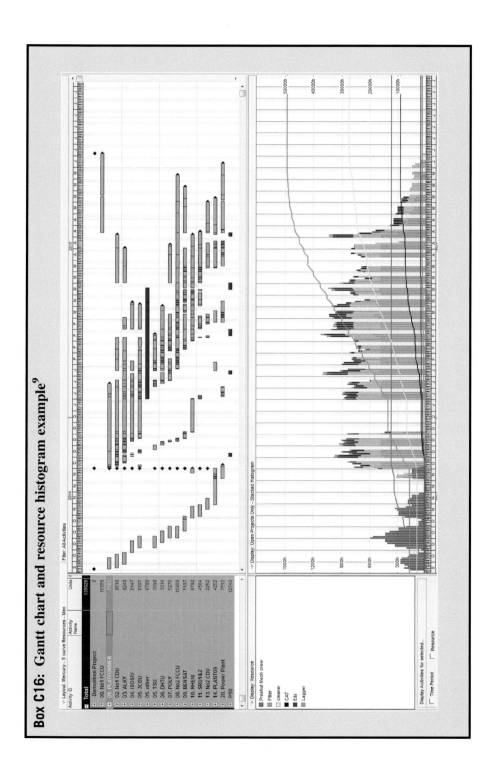

Box C17: Resource histogram example[9]

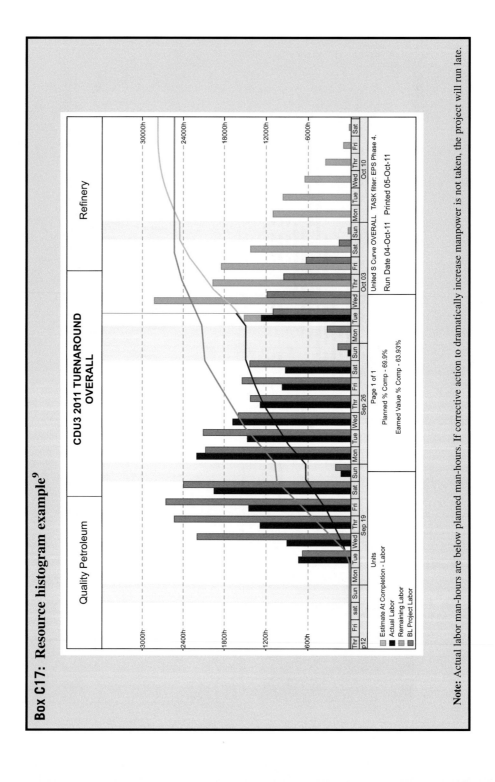

Note: Actual labor man-hours are below planned man-hours. If corrective action is not taken, the project will run late.

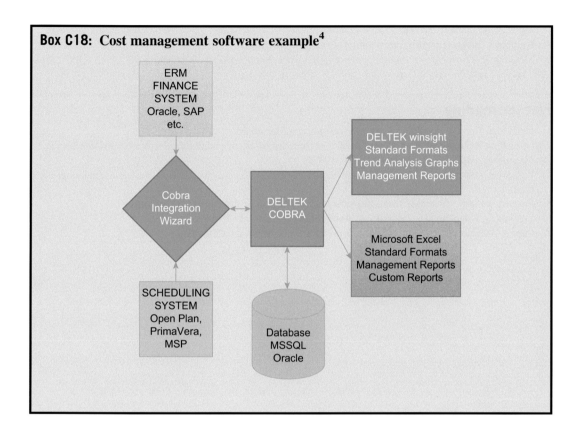

Box C18: Cost management software example[4]

- ERM FINANCE SYSTEM Oracle, SAP etc.
- Cobra Integration Wizard
- DELTEK COBRA
- DELTEK winsight Standard Formats Trend Analysis Graphs Management Reports
- Microsoft Excel Standard Formats Management Reports Custom Reports
- SCHEDULING SYSTEM Open Plan, PrimaVera, MSP
- Database MSSQL Oracle

References

1. PrimaVera P6 Resource Levelling. Available from: https://tensix.com/2013/02/resource-leveling-with-primavera-p6-professional/.
2. Primavera Risk Analysis. Available from: https://www.oracle.com/applications/primavera/products/risk-analysis.html.
3. Deltek Acumen. Available from: https://www.deltek.com/en-gb/products/project-and-portfolio-management/acumen.
4. Deltek Cobra. Available from: www.deltek.com/~/media/product%20sheets/cobra/deltek%20cobra%20product%20sheet.ashx?la=en-gb.
5. Navitrack. Available from: http://www.ap-networks.com/blog/shell-adopts-navitrack/.
6. IAMtech (developed in conjunction with BP). Available from: www.iamtech.com/shutdownturnaroundoutagesoftware.php?gclid=CjwKCAjw1tDaBRAMEiwA0rYbSPQ2ABc5PRqiCZRYNgtPPSXCtO_VgQZgcAFJPAPiD7il_ExN938UvxoCjc8QAvD_BwE&gclid=CjwKCAjw1tDaBRAMEiwA0rYbSPQ2ABc5PRqiCZRYNgtPPSXCtO_VgQZgcAFJPAPiD7il_ExN938UvxoCjc8QAvD_BwE.

7. Mpower. Available from: http://www.monitor-mpower.com/software/mpower/shutdown-and-turnaround/.
8. Kreiling J. Seven refineries, one solution. *Primavera Magazine*.
9. McGrath R. *The recipe for success: investing in comprehensive shutdown planning STO Asia*. IQPC; 2014.

Further reading

1. Open Plan website. Available from: www.deltek.com/openplan.
2. Open Plan Deltek Open Plan Demo. Available from: https://www.youtube.com/watch?v=nHNoujLiqjY.
3. *Open Plan professional user guide*. WST; 1997.
4. Primavera website oracle. Available from: https://www.oracle.com/industries/construction-engineering/index.html.
5. Primavera Demo. Available from: https://www.youtube.com/watch?v=zEL553w_snM.
6. Fleming Q, Koppelman J. *Earned value project management*. PMI; 1996.
7. Impact. Available from: http://www.impact-software.co.uk/Product/Planning.

Appendix D: New plant commissioning and start-up

D0. Overview

> I felt exactly how you would feel if you were getting ready to launch and knew you were sitting on top of 2 million parts—all built by the lowest bidder on a government contract.
>
> **Attributed to astronaut and US senator John Glenn.**

Objective

To bring the new plant online safely and in an acceptable time frame, so as to achieve design production rate and quality.

Note: When bringing new plant online, there will be a buffering period until the operation is stable enough to achieve the desired production rate and specifications.

Challenge

Conflicting objectives need to be aligned where possible. For example, contractors want to complete work and hand over as soon as possible, whereas commissioning personnel want to ensure the plant works as designed. Ideally, both commissioning personnel and contractors wish to meet the deadlines of mechanical certification and start-up.

D1. Phases
Planning: activities

Planning entails the development of extensive commissioning check lists, writing start-up, operating, and maintenance procedures, and training of operations and maintenance staff, all prior to commissioning and start-up. Details are discussed in Section D2: Knowledge areas.

Execution: sequence of events
Precommissioning

Precommissioning activities are the nonoperating work responsibilities, such as adjustments, cold alignment checks, etc., performed by the contractor prior to the "ready for commissioning" milestone or mechanical completion. These include original equipment manufacturer requirements.

Mechanical completion

Mechanical completion occurs when the plant, or any part of the plant, has been erected in accordance with drawings, specifications, and applicable codes, and the precommissioning activities have been completed to the extent necessary for the client to accept the plant and begin commissioning activities.

The client carries out numerous checks based on extensive checklists and produces a "punch" or "but" list of items to be rectified before a completion certificate is signed. Once this has been endorsed, the client is then ready to commission the plant.

Commissioning

Commissioning activities are associated with preparing for operating the plant or any part of the plant prior to the initial start-up and are usually the owner's responsibility.

Flushing and dry out

In most cases water has been used for hydro-testing. All water needs to be removed from the systems before product is introduced. In some cases the system is purged with steam or nitrogen.

Dry run (mock operation)

These are commissioning activities to test controls and rotating equipment.

Initial start-up

Initial start-up occurs when feedstocks are introduced to the plant. In refining operations, product is normally circulated before heat is introduced in the furnaces.

Product on spec

After start-up, numerous laboratory tests are carried out until the product meets specification for sale. Alternatively, online analyzers can be used to reduce laboratory tests. During this time extensive troubleshooting is anticipated. Once the design rate has been stabilized, performance tests are carried out. These often entail a contractual performance guarantee from the contractor, and plant acceptance follows once the performance tests have been approved by the client.

The above is outlined in Fig. D1.

On occasion, the plant is not able to reach the design performance parameters. If a design rate is guaranteed by the contractor, penalties may be applied, although this would depend on the feedstock being the same as that used for the basis of design.

Example:

A sample of raw gas used for the design of a gas plant may have a different composition to that which is eventually fed to the gas plant.

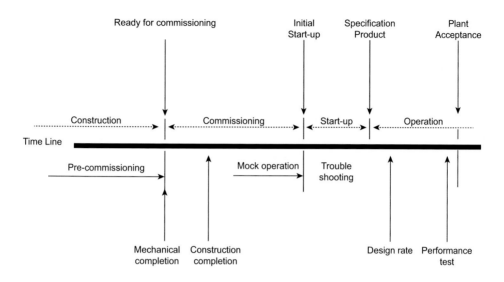

FIGURE D1

Commissioning and start-up of new plant.

D2. Knowledge areas

New plant commissioning and start-up is discussed under the PMBOK/ISO knowledge areas.

1. Integration
Activities
Management

Initial management activities include:
a) Confirm overall project timing
b) Develop a project control philosophy and tie it in with the turnaround philosophy
c) Establish staffing levels for the experienced staff that will be required
d) Develop an accurate estimate of start-up cost
e) Develop reasonable schedules and tie them in with turnaround dates
f) Develop personnel structures and responsibilities.

Operations

Initial operations activities include:
a) Define operational management boundaries
b) Agree on personnel responsibilities
c) Agree an overall time schedule that is compatible with the turnaround dates

d) Develop detailed recruiting schedules and training programs
e) Verify start-up cost estimates
f) Produce/review general checklists.

Parallel preparation process

Operations preparation needs to run parallel to the project progress so as to converge during commissioning and start-up. The client's project and operations teams need to work closely together through the precommissioning and commissioning stages. Separation of duties between the project team and the operating team is highly recommended.

An example is given in Box D1.

2. Scope

Scope is based on the design of the plant and is thus well defined. However, issues that arise when troubleshooting during start-up are not always foreseen. If new technology is being applied, troubleshooting may be extensive.

For all changes to scope during the construction phase, the management of change (MoC) process must be strictly adhered to, with relevant "as-built" piping and instrument diagrams (P&IDs) being produced for the final plant walk-down before start-up. A fast-track process for MoC may be necessary, but this must not compromise the basic MoC principles.

3. Schedule
Planning levels
Level 1: overall project schedule
a) Unit-by-unit start-up sequence
b) Engineering sequence involving project and plant management
c) Procurement sequence
d) Construction sequence.

Box D1: Petronas RAPID—scope of project team versus operations team

Precommissioning *Project team*	Commissioning *Operations team*
Walk-down, punch listing, and closeout	Mechanical run test, MV/HV motor solo run
System and piping pressure testing—hydro-test and pneumatic test	Inerting purge in/out, leak test, and pressurization
Flushing, blowing, and cleaning	Dynamic functional test/system acceptance test
Loop check, calibration, functional test, and site acceptance tests (SATs)	Furnace refractory dry out, catalyst, and chemical loading
LV motor solo run, electrical equipment and network commissioning and testing	Readiness for crude in/gas in and other feedstocks and materials
Equipment inspection, box-up, and preservation	Readiness for performance tests

Level 2: overall operations schedule
a) Detailed checklist schedule developed/agreed by operations management
b) Manpower buildup schedule.

Level 3: mechanical completion and commissioning schedule
a) Commissioning and start-up sequence
b) Hydro-static test sequence: start-up and operations supervision with construction and project input
c) Equipment sequence
d) Operational systems priority
e) Instrument checkout sequence.

Level 4: initial start-up schedule
a) Utility start-up sequence
b) Inventory sequence—plant operations, maintenance, and technical personnel with start-up personnel assistance
c) Equipment "on-circulation" or "reflux"
d) Feedstock introduction sequence (including possible management of simultaneous operations).

Planning tools

Planning tools include:
a) Bar charts
b) Network diagrams
c) Flowcharts.

Commissioning activities

General commissioning activities are shown in Box D2.

Box D2: General commissioning activities

1. Guarantee proper lubrication
    a) Check lubrication specification list against vendor recommendations and service requirements
    b) Review actual field lubrication supplies for comparison to specifications
    c) Follow up on actual in-plant lubrication program: Who does what and when?
2. Valves
    a) Lubricate valve stems and open and close every valve several times
    b) Check packing for tightness
3. Review catalyst and chemical status
    a) Analyze chemical and catalyst supplies, source, and back-up availability
    b) Arrange sufficient storage locations and adequate protection
    c) Check specification compliance
4. Conduct blind flange removal inspections
    a) Develop blank insertion diagrams
    b) Inspect lines before installation
    c) Keep track of blanks

Continued

Box D2: General commissioning activities—cont'd

5. Dry run
   a) Observe operator knowledge of equipment and valve locations
   b) Identify need for double operator coverage
   c) Identify need for operations supervisory, maintenance, and engineering coverage
   d) Identify additional instrumentation and tools for start-up conditions and troubleshooting
6. Instrument setting
   a) Review instrument settings relative to data sheets
   b) Scrutinize all alarms to determine their relative importance and urgency
   c) Establish prioritization hierarchies to assist the operator to manage an emergency situation
7. Identify piping and equipment
   a) Use color-coded paint/tape for arrows, signs, etc. to show material and flow direction. Use same color system on plant model and central control room graphics
   b) Stencil equipment numbers on vessels, towers, pumps, etc.
   c) Install valve tag numbers
8. Test flow systems
   a) Inspect sequence and clean out construction debris
   b) Conduct flow tests
9. Valve inspection
   a) Visually inspect blowdown valves to be sure they are open, then verify they can be closed
   b) Visually inspect double block and bleed, close blocks and open bleeds
   c) Visually inspect that all bypasses are closed and all block valves are open around control valves
10. Check suction screens
    a) Are they in?
    b) Do they stay in?
    c) Are they clean?
    d) Can one measure the pressure differential?
11. Check online filters
    a) For correct elements
    b) For proper installation
    c) Filters are in good condition after commissioning
12. Conduct final review
    a) Safety procedures
    b) Equipment clearing procedures
    c) Emergency reaction procedures
    d) Fire protection equipment
    e) Safety equipment
13. Reevaluate start-up manpower plans
    a) Sufficient in-house operating and maintenance staff
    b) Adequate technical and laboratory support
    c) Vendor requirements
    d) Specialist consultants
14. Update feedstock delivery program and product disposal plans
    a) Review tankage and availability
    b) Review delivery plans
    c) Check product shipment schedule
15. Assess environmental contingency plans
    a) Develop additional prestart-up governmental liaison as required
    b) Update contingency plans and who in what agency to contact

---

Box D2: General commissioning activities—cont'd

16. Review off-plot service facilities
    a) Contract maintenance
    b) Equipment cleaning
    c) Off-specification material disposal
    d) Scrap handling
    e) Workshop facilities
17. Perform motor run tests
    a) Check direction is correct
    b) Conduct run-in of uncoupled motors
    c) Check bearing noise and temperature (baselines)
18. Steaming out
    a) Develop criteria for cleanliness
    b) Log condensate trap inspection program
    c) Monitor steam line growth and movement
    d) Check supports
    e) Conduct testing at above design capacity
19. Chemical cleaning
    a) Establish what has to be cleaned and the extent of cleaning
    b) Determine suitable cleaning methods and contractors
    c) Set cleanliness inspection standards
    d) Monitor the cleaning process carefully
20. Line flushing
    a) Determine the flushing method, break points, and disposal
    b) Determine maintenance crew requirements
    c) Create a commissioning safety procedure
    d) Establish cleanliness standards.

---

4. Cost
Start-up cost definition

This is defined as all those noncapitalized costs and expenses incurred by the owner and associated with the overall plant operation from project inception through start-up and operation until some predefined production goal is attained. These costs are separate from, and in addition to, project costs and operating expenses. Significant items include off-specification products, manpower, and utilities.

Start-up budget preparation

Extensive overtime is required during commissioning, as commissioning staff work inordinate hours to stabilize the plant operation. This needs to be catered for in the budget.

5. Quality
Construction inspection

Construction inspection by the client's representatives is essential to ensure that documented specifications and design requirements are met. Inspections ensure the following:

I. Installation is as per design and specifications
II. Quality of workmanship meets industry standards

III. Design, vendor, and field mistakes are corrected
IV. Long-term integrity of the plant is ensured
 V. Start-up problems are minimized.

Product

Additional laboratory services are required as there is an increased frequency of sampling until the plant reaches a steady state.

Pressure-containing equipment and piping

Records for all pressure-containing equipment and piping are to be completed and handed over to the owner. These include material composition, welding, heat treatment, and pressure test certificates for manufacture and installation of pressure containing equipment and piping.

Instruments and controls

Acceptance tests are required:
1. Factory acceptance tests (FATs) are essential for new control systems and switch-gear
2. Site acceptance tests (SATs) then only deal with site installation related issues.

Electrical

FATs and SATs are required for electrical switch-gear. In certain cases, load testing of main power systems is required. This may entail the lease of resistive and reactive load-banks.[1]

Critical operating parameters

Initial set-up of operating limits is required at start-up. The safe operating limits (SOLs) should be defined in the safe operating envelope. These include:
1) Product quality limits (upper and lower)
2) Equipment integrity limits (ensuring operation within design limits)
3) Reliability limits (ensuring the correct predictive and planned maintenance is carried out)
4) Environmental permit limits (not to exceed license to operate limits)
5) Other operating limits.

Specific operating targets need to be determined and validated. These include:
- Product quality targets: determined by industry standards
- Optimization targets: determined by optimization of the process (e.g., APC)
- Throughput and yield targets: determined by design of the plant, simulation, and bench-marking
- Reliability targets: determined by the production cycle (see Section 2.3)
- Energy targets: determined by regulatory authorities and bench-marking
- Chemical/catalyst targets: determined by process operational severity and chemical/catalyst cost.

6. Resources
Start-up team

The start-up team includes the commissioning team leader, process engineers, process safety engineers, health, safety, and environment (HSE) engineers, software engineers, instrument engineers, electrical engineers, mechanical engineers and operating personnel. The team must have access to and priority over maintenance personnel and other necessary support. The team leader must have had, and most of the team should have had, prior commissioning and start-up experience.

Team office activities:
1. Assist contractor/owner in design reviews and P&ID screening
2. Help with the preparation of operating manuals
3. Help with the preparation of analytical procedures (product quality) manual
4. Provide manpower estimate and schedule for field start-up organization
5. Prepare a cost estimate of preoperational and start-up expenses
6. Help to prepare owner's organization charts and staffing requirements
7. Conduct and review risk assessment using various tools, i.e., HAZOP, etc.

Team field activities:
8. Review P&IDs
9. Oversee hydro-static testing
10. Conduct punch listing
11. Perform vessel inspection
12. Advise on flushing, steam blowing, and cleaning operations
13. Oversee instrument installation, calibration and loop testing
14. Prepare start-up activities schedule
15. Plan and coordinate vendor equipment run-in
16. Issue plant commissioning procedures
17. Help develop plant performance acceptance test procedures.

Resource levels

Common practice is to double up on operating resources during start-up by pairing experienced operators from other plants or affiliates, where possible, with inexperienced new operators.

Maintenance staff support should be at full strength during commissioning and work with the contractors to ensure a smooth handover.

Training
Typical training programs

Training for staff of a new plant includes:

Management and general
1. Management and supervisory techniques
2. Orientation

3. Language
4. Instructor training and teaching techniques

Operations
5. Process training of supervisory and technical personnel
6. Basic vocational training for inexperienced operators
7. Process training for experienced operators
8. Simulator training

Maintenance
9. Basic vocational training for inexperienced craftsmen
10. Advanced training for experienced craftsmen
11. Maintenance of specialized equipment for qualified craftsmen
12. Instrument technician training
13. Maintenance procedures and systems for supervisory and technical personnel
14. Planning and scheduling techniques and use of software

Laboratory
15. Laboratory technician training

HSE
16. Training of dedicated HSE staff
17. Preactivity safety review training for start-up staff
18. General HSE training: working at heights, defensive driving, confined space entry, etc.
19. Other HSE training as appropriate.

Operator recruitment and training

Common practice for major new installations is to send key operators to a similar plant for as much as a year for real operator experience prior to returning to site to commence commissioning and start-up of the new plant.

> *Example*
>
> *Chevron sent Thai operators to their Cape Town refinery for training for 12 months before they then returned to Thailand to commission and start up their new refinery in Map Ta Phut.*

Convergence between the recruitment and training of operating staff and construction occurs at construction completion and the start of commissioning. Complex plants require all panel operators to undergo simulator training.

7. Communication
Documentation

Extensive documentation, including numerous checklists, is required. Examples are shown in Box D3.

Box D3: Documentation examples

a) Organization and staffing
b) Recruiting plans and schedule
c) Training programs
d) Operating manuals and procedures
e) Administrative policies and procedures
f) Maintenance facilities
g) Maintenance tools and equipment
h) Operating supplies
i) Spare parts and consumables
j) Catalysts and chemicals
k) Safety and fire protection
l) Precommissioning
m) Commissioning
n) Operation
o) Performance testing
p) Detailed emergence procedures
q) Laboratory services
r) Start-up budget
s) Supplementary personnel
t) Support services.

Construction completion and inspection

Extensive documentation is required to be signed off. These include:
- Hydro-test certificates
- Loop tests
- Trip and alarm tests
- Completion of punch lists.

Commissioning

A lot of "hand holding" occurs at this time. Liaison between contractors, project managers, commissioning staff, maintenance staff, and operators is essential. Commissioning activities are listed in Box D2.

Start-up

Operations take over the plant for start-up and integration with existing plant operations.

Regular and frequent start-up meetings are required to coordinate activities. Agenda items include progress on the punch lists, extra/additional work, achievement of start-up dates, demobilization timing, and resource level reductions.

Some pointers for preparation of the start-up.
1. Communications: Hardware, radios, phones, Wi-Fi, public address. Are they adequate and reliable?
2. Line of command: Who issues instructions and how?

3. Access: Who is allowed at the control room panel? Who is not allowed at the control room panel? How is control room access going to be limited?
4. Manning: Long hours, call-ins, meals, transport
5. Holding patterns: Who decides? Forward momentum: How to maintain it?
6. Instrumentation: Tuning, changes to set points
7. Meetings: Avoidance, limiting duration, and restricting attendance
8. Logging: Process data, variances, changes and the decision processes
9. Process stability: Increasing and reducing production rates
10. Experience: How to apply it at the correct stages in start-up?
11. Troubleshooting: Who and how?
12. Problems: Levels and assignment for resolution
13. Maintenance paperwork: Tracking work done in start-up
14. Safety: Safe practices and procedures. Can design limits be exceeded? Should they? If so, in what instances?
15. Trust: Working relationships, knowing an individual's limits.
16. Unusual occurrences: Reports, historical analysis
17. Test runs: Preparation, guarantee testing
18. Assurance: Internal and external audit.

Procedure preparation[2]

The sequence for preparation of procedures is:
a) Generate a list of required procedures
b) Define who, what, where, and when for development
c) Develop in appropriate standard integrated management format in line with relevant ISO standards (9000, 14000, 45001, 31000, 55000).

Actual start-up needs to be checked for compliance with the written start-up procedures. These procedures may need updating after the start-up.

There should be a database to store all procedures or relevant process information so that the most updated version is readily accessible by all parties. Procedures should be reviewed and revalidated periodically.

Transfer of ownership

This change from the contractor to the owner requires extensive communication, both written and verbal. Initially there would be a two-way communication between the contractors and the project team. As completion draws nearer, the commissioning team starts to interface more actively with the project team. The commissioning team takes ownership from the contractor on behalf of the operator and then, once a steady state of production has been achieved, hands the operating plant to the operator.

8. Risk/HSE

Initial start-up is a particularly hazardous event. It is thus essential to embed a **corporate safety culture[3] throughout the organization** for safe commissioning and start-up.

Other risk/HSE issues are discussed in the following paragraphs.

Emergency planning

Emergency planning should be carried out as follows:
1) Identify possible emergency situations
2) Examine the possible consequences of each situation
3) Detail the actions, both automatic and otherwise, that should or could be taken
4) List the automatic systems, the equipment, and the resources required for each action
5) Ensure that these are, or will be, provided
6) Expect the unexpected
7) Review what has happened elsewhere on similar plants.

Examples of situations requiring emergency response plans are described in Box D4.

The above is part of incident preparedness and operational continuity management, which is regarded as best practice for all process plants.[4]

Safety problem minimization

Ways to minimize safety concerns:
1) Prepare normal safety procedures, start-up procedures, operating manuals, safety tests, and commissioning procedures well before commissioning and start-up.
2) Develop and dry-run emergency reaction plans.
3) Identify and educate staff about toxicity possibilities.
4) Maximize fire control availability.
5) Design to minimize loss of instrumentation.

Box D4: Emergency response situations

a) Power failure
b) Loss of instrument air
c) Loss of cooling water
d) Toxic gas release
e) Major leak
f) Spurious interlock functions
g) Fire
h) Spills
i) Runaway reactions
j) Loss of fuel supply
k) Loss of steam or boiler feed-water
l) Pump or compressor seal blowout
m) Over firing or tube rupture in fired equipment
n) Single-line equipment shutdown
o) Sewer explosion
p) Flare header rupture
q) Flare pilot failure.

---

6) Thoroughly study varying start-up conditions for controllability, unstable process conditions, and excessive temperature, pressure, or flows.
7) Review utility supply for start-up conditions.
8) Recheck fail-safe functions just before start-up.
9) Establish operational performance measurement, i.e., the reporting on the conformance to SOL excursion.
10) Limit accumulation of small problems with rapid resolution processes
11) Avoid starting long duration start-up activities with tired people by utilizing "holding patterns" to recover from fatigue
12) Encourage teamwork through training to prevent individual risk-taking
13) Conduct safety training/workshops to obtain appropriate certifications—hazards of nitrogen, chlorine, hydrogen sulfide, etc.
14) Adhere closely to the master checklist.
15) Plan for, and develop, shutdown procedures (both emergency and routine) well before commissioning and start-up.

Items listed above must be covered using one or more of the following risk tools.

Risk tools

Risk tools should include:
  I. Total review of all properties of all chemicals, catalysts, and hydrocarbons for toxicity, flammability, corrosiveness, etc.
 II. Review start-up process conditions for possible temporary generation of toxic or explosive mixtures
III. Identify difficult equipment purging or clearing possibilities and preplan for this
IV. Using chemical reactivity matrices and safety data sheets, etc., identify and educate about known problem sources such as pyrophoric reactions, hydrogen sulfide, high-temperature hydrogen, high-pressure steam, cyclic temperature or pressure operations, flame failure of fired equipment, and excessive start-up pressures, temperatures, power, etc.
 V. Check special safety equipment and clothing requirements.

Common risk tools are listed in Table D1.

Additional risk mitigation documentation includes:
1. Cause and Effect Matrix
2. Fire Safety Design Philosophy
3. Relief System Design Philosophy
4. Safeguarding Memorandum and Process Safeguarding Flow Diagram
5. Safety Equipment Layout and Escape Route
6. General Plant/Platform Alarm System Layout
7. Firefighting System Layout Drawing.

Table D1 Common risk tools.		
Acronym	**Description**	**Use**
FIREPRAN	Fire protection analyses	Project
FMEA	Failure mode effects analysis	Project
FEA	Finite element analysis	Project
EERA	Evacuation, escape, & rescue analysis	Project (offshore)
ESSA	Emergency system survivability analysis	Project
EIA	Environmental impact assessment	Project
SIA	Safety impact assessment	Project
FMECA	Fault modes, effects, & criticality analysis	Project
HSECES	Health, safety, & environment critical equipment & systems	Project and change management
HAZID	Hazard identification study	Project and change management
LOCPA	Loss of containment protection analysis	Project and change management
HAZOP	Hazard and operability study	Project and change management
LOPA/IPF/ SIL	Layers of protection analysis/instrument protective functions/safety integrity levels	Project and change management
PHA	**Process hazard analysis**	**Project and change management**
QRA	Quantitative risk analysis/assessment	Project and change management
HEMP	Hazards & effects management process (a Shell process)	Project and change management
RCA	Root cause analysis (Tri-beta & Taproot are common products used in the oil & gas industry)	Incidents
RCM	Reliability-centered maintenance	Operations
RBI	Risk-based inspection	Operations
SAFOP	Safe operations	Operations
	CIMAH/COMAH/demonstration of safe operation/HSE case	Project
	Blast study	Project
	Fire and explosion risk assessment	Project
	Flare radiation study	Project
	Gas dispersion study	Project
	IPF study	Project
	Fire & gas detection mapping	Project
	Fire water demand and hydraulic study	Project
	Acoustic induce vibration	Project
	Occupied building risk assessment	Project

Prestart-up safety reviews

The purpose of a prestart-up review (PSSR) is to verify that all elements required for safe operation of the process have been addressed prior to start-up. This ensures that construction is in accordance with design specifications, all necessary procedures for safe operation are completed, applicable process safety information is up-to-date, personnel responsible for operating and maintaining the plant are properly trained, and all recommendations made during a process hazard assessment (PHA) have been addressed.

Many catastrophic accidents occur during the start-up of new or modified plant or equipment. Prevention of accidents in the start-up stage is thus of prime importance. The goal of PSSR is to prevent these accidents by verifying that:
1. Process equipment and piping have been properly designed, constructed, and installed
2. Recommendations from PHA studies, other safety reviews, and accident reports have been addressed and implemented
3. Process safety information is accurate and complete
4. Emergency and operating procedures have been developed, and
5. Operators and other appropriate personnel have been trained in their new or modified job functions.

The PSSR is usually performed by a team whose members are experts in the functions mentioned. Review includes both paper and physical inspection of the process equipment, piping, and instrumentation to verify that the plant was built in accordance with the design and is ready for start-up. In addition, the procedures for performing the start-up are reviewed to ensure that each action is performed safely. A PSSR checklist comprises checklists for different disciplines and is tailored for different units and purposes, e.g., a PSSR checklist for process start-up should be slightly different from an off-site start-up.

Performing a PSSR is critical for safe operations:
  I. after a turnaround for maintenance and inspection, or
 II. after major alterations to the process, or
III. once installation of new plant and equipment has been completed.

Cases I and II are an integral part of the MoC process. To ensure safe operations, PSSR must also be conducted prior to introducing hazardous chemical substances or feed to the process units. Box D5 outlines a suitable template for PSSR. For maintenance and inspection turnarounds, a more simplified template is shown in Box 7.9.

Contingency planning

Stock materials for a new process plant may not be available at the time of start-up. Therefore, start-up materials need to be made available and kept on hand. Bear in mind that items such as filters and gaskets are consumed at a higher rate during start-up.

An abnormal operation or plant safe operating limit (SOL) excursion, or even a unit operation shutdown, may necessitate maintenance support and/or additional stock materials.

P&ID walk-through

The engineering and operations staff need to walk through and review the plant against the "as-built" P&IDs as part of the PSSR. This is a legal requirement in certain countries, and any deviation of actual layout from the P&ID could thus have legal implications.

Box D5: PSSR template

1. Purpose

This procedure details the application, responsibilities, and management systems required for implementation of the PSSR. The purpose of a PSSR is to verify that all elements required for safe operation of the plant have been addressed prior to start-up, ensuring that:

A. Construction is in accordance with design specifications
B. All procedures necessary for safe operation are completed
C. Personnel responsible for operating and maintaining the plant have been properly trained
D. Applicable process safety information is up-to-date, and
E. All recommendations made during process hazards analysis have been addressed and completed as appropriate.

2. Scope

The PSSR procedure applies to:

1) New construction
2) Major modifications to existing facilities, and
3) Major MoC projects.

3. Definitions

Major modification: Multiple simultaneous hardware changes to a plant designed to update, revise, or reconfigure the process technology. Major modifications are significant capital projects usually requiring substantial unit downtime to perform the work.

Post start-up action items: Deficiencies found during the PSSR that will not result in a hazard if the plant is started up.

Prestart-up action items: Deficiencies found during the PSSR that may result in a hazard if the plant is started up and that must be corrected prior to start-up.

PSSR: Prestart-up safety review.

4. References

Implementation of this process will be in accordance with other process safety management processes as appropriate, in particular the following:

   I. Process safety information
  II. Process hazard analysis
 III. Operating procedures
  IV. Training
   V. Mechanical integrity
  VI. Management of change
 VII. Safe work practices
VIII. Emergency planning and response

5. Authority and exceptions

The operating superintendent or designated representative is authorized to approve start-up of the plant after completion of the PSSR and resolution of all prestart-up action items.

Other plant personnel are authorized to perform their duties as outlined in the responsibilities section of this procedure.

6. Responsibilities

Operating superintendent
- Attends PSSR planning meeting
- Attends PSSR resolution meeting
- Reviews resolution of all prestart-up action items, and
- Approves start-up of the plant following completion of all prestart-up action items.

Project engineer
- Acts as PSSR team leader
- Conducts PSSR planning and resolution meetings
- Assigns PSSR team member tasks and schedule

Continued

Box D5: PSSR template—cont'd

- Tracks resolution of all action items to closure, and
- Issues memorandum documenting PSSR completion.

Process safety management coordinator
- Coordinates development of PSSR checklists
- Assists training manager in developing employee training on PSSR, and
- Maintains documentation of completed PSSRs in master process hazard analysis file.

Training manager
- Ensures all employees receive PSSR training as appropriate.

Employees
- Participate as PSSR team members as required.

7. Requirements

General

Note: It is not practical to perform a formal PSSR on every modification handled via the MoC process. PSSRs should be used in complex construction projects where the likelihood that some aspect of installation will not conform to design is fairly high.

The process safety management coordinator shall work with appropriate plant personnel to develop a standard PSSR checklist. The checklist shall include verification that:

1) Process safety information is complete
2) Construction is in accordance with design specifications
3) All procedures, including operating, maintenance, emergency, and safety, necessary for safe start-up and operation, have been prepared and communicated to employees
4) Appropriate training has been conducted
5) Appropriate equipment test and inspection procedures and frequencies have been established
6) All other necessary systems are in place, and
7) The plant emergency response plan is still adequate.

The PSM coordinator shall work with the training manager to develop employee training on PSSRs.

The training shall differentiate between MoC and PSSR requirements.

PSSR team

The project engineer is designated as the PSSR team leader and shall select other team members and delegate responsibilities for conducting the PSSR.

The PSSR team shall consist of persons from the following disciplines:

1) Process engineering
2) Mechanical inspection engineering
3) Safety
4) Environment
5) Operations
6) Maintenance.

Planning the PSSR

The PSSR team leader shall conduct a preview meeting with the PSSR team and the appropriate operating superintendent.
- The team leader shall present an overview of the project for the benefit of the PSSR team members
- The PSSR checklist shall be reviewed and modified as appropriate
- The team leader shall assign portions of the PSSR to appropriate team members
- A completion schedule for the PSSR shall be developed.

Conducting the PSSR

PSSR team members shall evaluate the plant using the PSSR checklist agreed upon during the PSSR planning meeting.

Evaluation shall include a review of applicable documentation and field reviews of installations.

Deficiencies relative to the checklist shall be documented in writing by the reviewing team member.

Box D5: PSSR template—cont'd

Resolution meeting

Following the completion of the PSSR checklists, the PSSR team will meet with the operating superintendent to review the findings of the team.

A list of specific deficiencies to be addressed shall be developed during the resolution meeting. Deficiencies will be categorized as follows:

1) Prestart-up action items
2) Poststart-up action items.

Responsibilities and schedules for completion of all action items shall be specified.

Follow-up

The project engineer shall track all the PSSR items to closure.

Upon completion of all prestart-up action items, the project engineer will review the resolution of each item with the operating superintendent. If all prestart-up action items have been completed to the operating superintendent's satisfaction, the plant will be approved for start-up.

Documentation

All PSSR documents, including checklists, reports, and action item resolutions, shall be filed in the project file.

Upon resolution of all action items, the project engineer shall issue a memorandum to the operating superintendent and PSM coordinator that documents that the PSSR and all action items have been completed and specifies the location of the PSSR files.

The PSM coordinator shall file the memorandum with the master copy of the process hazard analysis for future reference.

8. Training

PSSR leader should be trained to meet certain requirements.

No specialized training is required for members of the PSSR team.

9. Audit

Each PSSR will be audited by the PSM coordinator for completeness and content. Deficiencies will be noted and assigned ownership, of which some may delay start-up of the new plant.

10. Attachments and examples

Attachment A: PSSR checklist for turnarounds.

Attachment B: PSSR checklist for new/altered plant and equipment.

Permit to work

The permit to work process switches from a construction environment to a live plant environment with the application of lock tag and try (LTT), and other isolation processes.

Potential risk issues and associated mitigation steps
Typical start-up complications

During commissioning, initial operation, and the first 6 months of operation, a high percentage of lost time (due to slippage, work holdups, unscheduled shutdowns, equipment outages, and plant downtime) can be attributed to:

Mechanical equipment failures
- Design—inadequate application of design standards (see Box D6)
- Fabrication and installation—inadequate application of procurement standards
- Inspection—inadequate application of inspection standards.

> ### Box D6: Third-party compliance checks—Petrobras Rig Collapse[5]
> #### A case of shortcutting standards
>
> In March 2001, a series of explosions sank Petrobras Platform 36.
>
> "The project successfully rejected … prescriptive engineering, onerous quality requirements, and outdated concepts of inspection …" A Petrobras executive, prior to the accident, on delivering superior financials.
>
> Proximate cause:
> - Leakage of volatile fluids burst a shutdown emergency drain tank and set off a violent chain of events.
>
> Underlying issues:
> - **A corporate focus on cost-cutting over safety**
> - Poor design of individual parts (with regard to system safety)
> - Component failure without sufficient backups
> - Lack of training and communication.

Faulty design
- Improper sizing
- Incorrect metallurgy
- Inadequate data (*example: incorrect feedstock specifications*)
- Scale-up errors—prototype to production (see Box D7).

Human errors
- Lack of communication
- Improper operation—lack of experience
- Poor maintenance practices—lack of training.

Software problems
- Inadequate testing—FAT and SAT
- Inadequate specification—design standards/design experience.

> ### Box D7: Investment risk—new technology
>
> Having major gas reserves, a national oil company decided to invest in a gas-to-liquid plant that involved new technology. It signed a joint venture agreement with a partner who had previously designed and operated similar plants.
>
> The processes/systems, people, and performance management aspects were all well established based on those of the venture technology partner. However, new technology involved in the process was based on a small-scale pilot plant and had not been tested in a full-scale commercial plant.
>
> Return on investment was based on a specific throughput (design capacity) and product "slate" price.
>
> After major project cost and time overruns, the plant finally started up but could only achieve half the design capacity. Nevertheless, the shareholders were delighted, as by a happy coincidence, the product "slate" price had doubled on the market.
>
> The plant eventually achieved design capacity, after more than a year of production.
>
> #### Comment
> Shareholders were aware of the risk of the new technology, but were still willing to take that risk.

Problem sources

Sources of problems could include:
1. Contract negotiations
2. Engineering contractor performance
3. Procurement specifications and pricing
4. Construction workmanship
5. Financial restraints
6. Planning deficiencies
7. Organizational weaknesses
8. Operating group performance.

Further pointers on reasons for difficulties outlined above:

1. Contract negotiations
a. Salesmanship by both parties: for example, a contractor overselling their services or a client trying to get the contractor to do more than they are capable of.
b. Nonadherence to the contracted schedule: monitoring methods and tools are essential.
c. Either party creating a win-lose situation: win-win situation is preferable.
d. High owner expectations: there has to be a meeting of minds as to what is expected.
e. Sidestepping the intent of the agreement for financial reasons—due to budget restraints.
f. Misunderstanding of contract language—verbal explanations clarifying legal language can be useful.
g. Inadequate process for dealing with unanticipated problems—a contract change management process is advised.
h. Inadequate definition of project scope—a wide bid range could indicate poor scope definition.
i. Inadequate definition of contractor responsibilities as a result of poor documentation or poor relationship.

2. Engineering contractor performance
a. Inexperienced staff
b. Work overload
c. Lack of people continuity
d. Inadequate estimate
e. Ineffective internal communication
f. Lack of flexibility when faced with exceptional situations
g. Schedule too tight
h. Too little delegation
i. New technology
j. Inadequate progress assessment.

3. Procurement specifications and pricing

a. Disagreement on acceptable vendor lists—an approved vendor selection process must be agreed upon
b. Reimbursement—client reneges on contractual obligation
c. Low-low bidder—contractor makes a loss to get the contract and then claims extras at a later time in order to make a profit, poor estimating
d. Split purchasing responsibilities between project management and contract management—a clear responsibility matrix is required
e. Late design changes
f. Delivery delays
g. Inadequate inspection program.

4. Construction workmanship

a. Inexperienced construction management and supervision—client should specify individuals with a proven track record
b. Nondedicated personnel
c. Lump sum subcontractor's optimistic schedules
d. Unable to achieve projected manpower buildup curve
e. Poor materials receiving and issuing procedures
f. Slippage of material and equipment delivery
g. Increased number of engineering changes
h. Inadequate inspection by unqualified personnel
i. Declining workmanship and morale.

5. Financial restraints

a. Optimistic cost justification
b. Tight budget
c. Low estimates
d. Departmentalized budgetary approach
e. Market changes—raw material and product prices.

6. Planning deficiencies

a. Too little, too late
b. Inadequate planning at every level
c. Insufficient detail
d. No regular updating with more accurate estimates
e. Lack of commitment to schedules
f. System becomes overly voluminous and complex
g. Poorly developed contingency plans.

7. Organizational weaknesses
a. Inadequate assessment of project staff requirements
b. Vague definition of engineering stage goals
c. Personnel responsibilities are not well defined
d. Inexperienced project engineering and construction personnel
e. Tendency to follow preliminary guidelines on cost and staffing
f. Failure to periodically appraise project and plant personnel
g. Lack of coordination between project and plant personnel
h. Holding an unrealistic schedule
i. Lack of appreciation for personnel continuity
j. Management unresponsiveness or interference
k. Lack of recognition of unique factors
l. Lack of coordination between construction and commissioning activities
m. Poor vendor support.

8. Operating group performance
a. Poor operational planning
b. Inadequate number of people
c. Inadequate training
d. Lack of recognition of unique factors
e. Management unresponsiveness or interference
f. Underestimate of start-up budget (separate from the project budget)
g. Noninvolvement in construction completion
h. Short duration construction
i. Poor operational decisions
j. Insufficient experience at key levels
k. High attrition rates.

In summary, problem sources could be grouped as follows:
1. **Inadequate management at project inception** as a result of poor experience, inadequate financial considerations, and/or poor planning.
2. **Poor execution of the project** due to inexperience, inadequate staffing, budget restrictions, weak leadership, and/or poor communication.
3. **Misoperation of the start-up** due to inexperience, inadequate commissioning, schedule pressures, financial pressures, and/or inadequate design.

Registers

Potential issues could be assessed for inclusion in a risk register so that mitigation action could take place to reduce the overall risk of a late start-up that is over budget. A MOC register and a deviation register also need to be maintained.

Incident-free operation

The application of process safety program tenets from initial start-up is highly recommended. These are:
1. Operate within limits and know your limits
2. Comply with all applicable rules and regulations
3. Address abnormal conditions via "triple S": stabilize, slowdown, shutdown
4. Ensure safety devices are in place and functioning
5. Follow safe work practices and procedures
6. Maintain integrity of dedicated systems
7. Use trend graphs to proactively monitor the most important process variables
8. Conduct operating rounds to look, listen, and feel
9. Meet customer's requirements/targets within the operating limits to avoid abnormal situations
10. Involve the right people in decisions that affect procedures and equipment.

9. Procurement
Commissioning spares

Setting up stock of spare parts for a new plant normally runs in parallel with commissioning and start-up, and spares are therefore not always immediately available. Consequently, spares such as mechanical seals, bearings, filters, etc., which are required for the most likely failures during start-up, need to be ordered and placed in stock prior to commissioning.

Support personnel

Additional operating staff and key specialist contractors need to be on standby during the start-up stage.

10. Stakeholder
Start-up management

The change of ownership requires the involvement of both the owner and operator. The operator is phased in with commissioning and start-up activities until a point is reached where ownership transfers from the contractor/s to the owner of the assets.

Required notifications

Regulatory authorities and insurance providers need to be contacted with respect to the change from project to operation mode and the date of transfer of ownership.

> *Example 1*
>
> *Environmental emissions are greater when a gas plant starts up, requiring relevant stakeholder notifications.*
>
> *Example 2*
>
> *Insurance changes from the contractor's construction insurance to the owner's asset insurance.*

> **Box D8: Control room mock-up**
>
> When designing the layout of a new control room, the Bahrain Petroleum Company built a complete full-scale mock-up of the proposed control room inside an old abandoned canteen that still had air-conditioning.
>
> Operators were able to get a "feel" of what to expect from the new control room. Many recommendations for an improved layout were proposed as a result.

Control room layout

Gaining acceptance of a new control room layout by operations staff is critical to the success of a project. An example is discussed in Box D8.

D3. Summary

Key elements of commissioning and start-up are outlined. Extensive planning is required to ensure a safe and trouble-free start-up and initial operation of a new plant. A PSSR is a key requirement prior to start-up.

References

1. Aggreko website. Available from: www.aggreko.com.
2. Sutton IS. *Writing operating procedures for process plant.* Chapman & Hall; 1995.
3. International Association of Oil and Gas Producers (IOGP) 20 elements of safety culture. Available from: http://www.iogp.org/.
4. Hey RB. IPOCM process. In: *Performance management for the oil, gas and process industries.* Elsevier; 2017. Available from: https://www.elsevier.com/books/performance-management-for-the-oil-gas-and-process-in-dustries/hey/978-0-12-810446-0. chapter 14.
5. Petrobras P36 rig collapse. Available from: https://sma.nasa.gov/docs/default-source/safety-messages/safe-tymessage-2008-10-01-lossofpetrobrasp36.pdf?sfvrsn=4.

Further reading

1. API 700 checklist for plant completion.
2. Lieberman NP. *Troubleshooting process operations.* PennWell; 1991.
3. Lees FP. *Loss prevention in the process industries.* 4th ed. Butterworth-Heinemann; 2012.
4. Horsey D. *Process plant commissioning.* 2nd ed. IChemE; 1998.

Appendix E: Framework documents

E0. Introduction

The framework documents should consist of two higher level documents: a turnaround management framework that applies to all turnarounds, and a turnaround manual that applies to each individual turnaround. Suitable contents for the turnaround manual could include the 10 knowledge areas based on the project management standards. It is advisable to ensure alignment with various ISO standards to ensure continuous improvement when compiling framework documentation. These are:

1. ISO 9000 Quality Management
2. ISO 55000 Asset Management
3. ISO 31000 Risk Management
4. ISO 14000 Environmental Management
5. ISO 45001 Occupational Health & Safety Management

ISO 9000 is a set of international standards applicable to quality management and quality assurance that have been developed to help companies effectively document the quality system elements that need to be implemented to maintain an efficient quality system. They are not specific to any one industry and can be applied to organizations of any size. ISO 9000 can help a company satisfy its customers, meet regulatory requirements, and achieve continual improvement. However, it should be considered to be a first step, the base level of a quality system, and not a complete guarantee of quality.

The ISO 9000 family contains these standards:

a. ISO 9001:2015: Quality management systems—Requirements
b. ISO 9004:2018: Quality management systems—Managing for the sustained success of an organization (continuous improvement)
c. ISO 19011:2018: Guidelines for auditing management systems.

ISO 55000 is an international standard covering management of physical assets.

The ISO 55000 family contains these standards:

a. ISO 55000:2014 Asset management—Overview, principles, and terminology
b. ISO 55001:2014 Asset management—Management Systems—Requirements
c. ISO 55002:2014 Guidelines for the application of ISO 55001.

ISO 31000 is a family of standards relating to risk management.

The ISO 31000 family contains these standards:

a. ISO 31000:2018—Principles and Guidelines on Implementation
b. ISO/IEC 31010:2009—Risk Management—Risk Assessment Techniques
c. ISO Guide 73:2009—Risk Management—Vocabulary.

ISO 14000 is a series of environmental management standards that provide a guideline or framework for organizations that need to systematize and improve their environmental management efforts.

The ISO 14000 family includes these standards:
a. ISO 14001:2015 Environmental management systems—Requirements with guidance for use
b. ISO 14004:2016 Environmental management systems—General guidelines on implementation.

ISO 45001 is an international standard for management systems of occupational safety and health, published in March 2018. The goal of ISO 45001 is the reduction of occupational injuries and diseases.

These standards have a common format to ensure continuous improvement.
1. Introduction
2. Scope
3. Normative references
4. Terms and definitions
5. Management system requirements
6. Management responsibility
7. Internal audits
8. Management review
9. Improvement.

E1. Turnaround management framework *(the document applicable to all turnarounds)*

Introduction

The value of effective turnaround management related to company profits.

Pictorial view

The turnaround management framework consists of all required documentation for managing all turnarounds and also includes a manual for managing specific turnarounds. Fig. E1 shows a possible outline.

Strategic planning of turnarounds

A statement on alignment with corporate direction.

Vision

Where we want to be.

A few simple points (see Box 2.1).

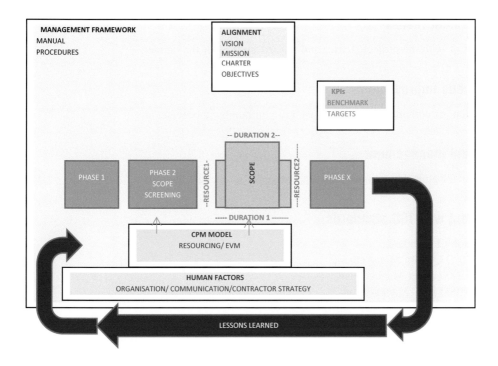

FIGURE E1

Turnaround management framework.

Mission

How do we plan to get there.

A few points (see Box 2.2).

Key performance indicators and targets
1. Long term
2. Specific turnarounds.

Benchmarking

A process for the benchmarking of each turnaround after completion.

Internal audits

Audits at the end of each phase.

Management review

Approval of both the project charter and phased updates of the budget.

Continuous improvement

A statement of commitment and approach to continuous improvement.

Document management

A process for embedding continuous improvement in documentation.

Alignment with ISO standards

Application of Standards:
  ISO 9000 Quality Management
  ISO 55000 Asset Management
  ISO 31000 Risk Management
  ISO 14000 Environmental Management
  ISO 45001 Occupational Health & Safety Management.

Organization and contracting philosophy

The approach to managing all turnarounds.

Managing the corporate knowledge base

  Inspection
  Maintenance
  Operations
  Asset Integrity
  Turnaround models.

E2. Turnaround manual *(applied to a particular turnaround)*
Contents

  1. Integration
     1.1. Initiating a turnaround
     1.2. Preparation of a project charter
     1.3. Drafting a project management plan
     1.4. Validating shutdown and start-up plans
     1.5. Managing organization process assets

2. Scope
    2.1. Formulating and validating a work list
    2.2. Controlling extra and additional work
    2.3. Scope challenge process
3. Schedule
    3.1. Writing job instruction packages
    3.2. Building a critical path model
    3.3. Critical path model software management
    3.4. Analyzing critical paths
    3.5. Archiving critical path models
4. Cost
    4.1. Formulating and controlling a turnaround budget
    4.2. Estimating
5. Quality
    5.1. Writing quality plans
    5.2. Welders' records
    5.3. Welding procedures
    5.4. Product quality procedures for plant start-up
    5.5. Key equipment method statements
6. Resources
    6.1. Creating turnaround teams and organizations
    6.2. Special crane lifts: lifting calculations and drawings
    6.3. Spares resourcing
    6.4. Tools and equipment resourcing
7. Communications
    7.1. Controlling documentation
    7.2. Controlling site logistics
    7.3. Daily routines during execution
    7.4. Problem solving
    7.5. Briefing and debriefing personnel
    7.6. Writing a final report
    7.7. Turnaround preparation checklist
    7.8. Turnaround start-up checklist
8. Risk/Health, safety, and environment
    8.1. Creating a safe work environment
    8.2. Emergency planning
    8.3. Conducting a site inspection and handover
    8.4. Developing and maintaining a risk register
    8.5. Turnaround environmental guidelines
    8.6. Permit to work
        8.6.1. Permit issuing procedure
        8.6.2. Confined space access procedure
        8.6.3. Confined space air testing procedure

Note: Relationships with other systems need to be identified in the relevant procedures.

Box E1 shows the contents of the Chevron Turnaround Planning and Execution Process. Note that it does not necessarily cover all aspects of the above framework.

Box E1: Chevron turnaround planning and execution process

Table of Contents
- Introduction
- Turnaround philosophy
- Turnaround planning process—Overview
- Plan development—Phase description
- Turnaround planning—Execution road map
- Appendices
  - Core team
  - Team commitment statement—Example
  - Shutdown philosophy—Example
  - Contracting strategy
  - Expenditure request process
  - Audit function
  - Criteria for shutdown work list items
  - Work list—Example
  - Milestone schedule for planning
  - Critical path scheduling
  - Shutdown execution
  - Additional work definition
  - Conflict resolution process
  - Final shutdown report guide

Adapted from Chevron Turnaround Planning and Execution Process (CTPEP).

Reference

1. Chevron Turnaround Planning and Execution Process (CTPEP).

Further reading

1. ISO 9000 Quality Management. Available from: https://www.iso.org/iso-9001-quality-management.html.
2. ISO 55000 Asset Management. Available from: https://www.iso.org/standard/55088.html.
3. ISO 31000 Risk Management. Available from: https://www.iso.org/iso-31000-risk-management.html.
4. ISO 14000 Environmental Management. Available from: https://www.iso.org/iso-14001-environmental-management.html.
5. ISO 45001 Occupational Health & Safety Management. Available from: https://www.iso.org/iso-45001-occupational-health-and-safety.html.
6. Sutton IS. *Writing operating procedures for process plant*. Chapman & Hall; 1995.

Appendix F: Further reading

1. Lenahan T. *Turnaround shutdown and outage management*. Elsevier; 2006.
2. McLay A. *Practical management for plant turnarounds*. 2003.
3. Levitt J. *Managing maintenance shutdowns and outages*. Industrial Press; 2004.
4. Sahoo T. *Process plants: shutdown and turnaround management*. CRC Press; 2013.
5. Brown V. *Managing shutdowns, turnarounds, and outages*. Audel; 2004.
6. TAR fighter shutdown project simulation. Available from: https://www.tacook.com/en/expertise/turnarounds-shutdowns-outages/.
7. Krings D. Proactive approach to shutdowns reduces potlatch maintenance costs. Available from: https://www.idcon.com/resource-library/articles/planning-and-scheduling/454-proactive-approach-shutdowns.html.
8. Motylenski J. Proven turnaround practices: maintenance and reliability. *Hydrocarbon Processing* 2003;**82**(4).
9. Elwerfalli A, Khan MK, Munive JE. A new methodology for improving TAM scheduling of oil and gas plants. In: *Proceedings of the world congress on engineering 2016*, vol. II. WCE; 2016. June 29–July 1, 2016, London, UK.

Appendix G: Glossary of terms

A

Accountability Total responsibility of an individual for the satisfactory completion of a specific assignment

Activity A task requiring resources, time, and budget

ACWP Actual Cost of Work Performed

ADNOC Abu Dhabi National Oil Company

AIM Asset Integrity Management

API American Petroleum Institute

ASME American Society of Mechanical Engineers

Audit Methodical examination to assess overall performance and compliance

Availability The percentage of the year that the plant is available to produce product

B

Baseline The plan against which performance is measured

BCWS Budgeted Cost of Work Scheduled

BCWP Budgeted Cost of Work Performed

Benchmarking Comparing performance to peers or leaders in order to provide a higher standard for the purposes of improvement

Bid Offer to perform the work described in a set of bid or tender documents at specified costs

BoQ Bill of Quantities. Used to price estimated quantities of items for bidding, but payment is made on actual quantities

BP&B Business Plan and Budget—submitted to the board of directors for approval on an annual basis

BPI Budget Proposal Item

C

CBS Cost Breakdown Structure

CM Corrective Maintenance

CMMS Computerized Maintenance Management System

Commitment For materials, it is cost of an item at the time of placing the order. For services, it could be a calculated labor cost based on time sheets. The intent of a commitment is for it to reflect the actual cost that the company is committed to pay up to today if, for instance, work stops now. This is used for cost tracking purposes as the actual payment could lag by a number of months. Financial reports would have budget, commitment, and expenditure columns. Performance reports would have budget, earned value, and actual cost (including commitment).

Constructability The optimum use of construction knowledge and experience in planning, engineering, procurement, and construction to achieve project objectives

Contingency A provision for those variations to the base estimate that are likely to occur, but cannot be specifically identified at the time the estimate is prepared (see also management reserve)

CPI Cost Performance Index

CPM Critical Path Method or Critical Path Model

Critical Path The series of activities in a network diagram that determine the earliest completion of a turnaround

CSFs Critical Success Factors

CV Cost Variance

D

Data Date See also Time Now—critical path model data entry and analysis date

DEP Design Engineering Practices

Digital twin A digital replica of physical assets, processes, people, places, systems, and devices that can be used for various purposes.

E

Efficiency KPI Measures the company's performance against similar companies

ER Expenditure Request

ERM Enterprise Resource Management—corporate wide management system. Leading software suppliers are SAP and Oracle.

EVM Earned Value Method (see Section A4.4 for definitions of terms used for EVM)

EWO Engineering Work Order—used for any change to the existing plant or process configuration

F

FBD Final Budget Decision

FCCU Fluidic Catalytic Cracking Unit

FEL Front End Loading—a quantitative measure of the level of definition for a turnaround (see Section 5.8.3)

FID Final Investment Decision

Fragnet PrimaVera term for a part of a project

FTA Fault Tree Analysis

G

Gantt chart A type of bar chart that illustrates a project schedule. This chart lists the tasks to be performed on the vertical axis, and time intervals on the horizontal axis. The width of the horizontal bars in the graph shows the duration of each activity.

GE General Electric Company

H

HAZAN Hazard Analysis

HAZOP Hazard Operability Study

Hot tap The cutting of a hole in a pipeline while the product is flowing without loss of product

HSE Health, Safety, and Environment

HSEQ Health, Safety, Environment, and Quality

I

IBD Initial Budget Decision

IID Initial Investment Decision

IOC International Oil Company

IPRT Independent Project Review Team

ISO International Standards Organization

ITB Invitation to Bid

J

Job Generic term for an activity or group of activities

K

KPIs Key Performance Indicators

L

LNG Liquefied Natural Gas
LTI Lost Time Injury

M

Management Reserve Funds not included in the turnaround budget that management may release for exceptional work or unknown situations that may occur during the turnaround execution phase.
MESC Materials and Equipment Standards and Codes
Milestone An activity with zero duration
MMP Major Maintenance Project
MoC Management of Change—A formal process for reviewing and approving any change to the existing plant or process.

N

NDT Nondestructive Testing
Not in kind Replacement of plant and equipment that is not identical physically or in material content. This is required to go through the management of change process.
NOC National Oil Company

O

OBS Organization Breakdown Structure
OEM Original Equipment Manufacturer
OOM Order of Magnitude estimate ($\pm30\%$)
Oracle American Enterprise Resource Management (ERM) software company

P

P&ID Piping and Instrument Diagram—a graphical representation of the process showing every process pipe, valve, vessel, and related instruments
PBD Preliminary Budget Decision
PERT Program Evaluation and Review Technique—uses a statistical technique to determine an activity duration
PET Polyethylene Terephthalate
PID Preliminary Investment Decision
PIN Project Initiation Note
Planned Shutdown A break in production that is planned for more than a year in advance (normally a separate budget item to maintenance management) (see also *Turnaround*)
Planning versus Scheduling **Planning** is the process of breaking down a turnaround into manageable work elements, shaping the work environment and identifying the resources required to efficiently perform the work. **Scheduling** is the process of defining the sequence and time frames in which each work element should be performed, and thus, determining the start and finish dates of the turnaround.
Proactive Maintenance Maintenance work executed to prevent failures or to identify defects that could lead to failure (failure finding). It includes routine preventive and predictive maintenance activities and work tasks derived from them.
PM Project Management

PM Preventive Maintenance
PM Predictive Maintenance
PM Planned Maintenance
PMBOK A Guide to Project Management Body of Knowledge
PPE Personal Protection Equipment
PPM Planned Preventive Maintenance
PSM Process Safety Management
PSP Process Safety Program
Punch List A list prepared when the project is almost complete to show just those items of work remaining to fulfill the project scope
Pyrophoric Chemicals Liquids, solids, and gases that ignite spontaneously in normal atmospheric conditions at or below 130°F/54°C.
PWHT Postweld Heat Treatment—required to stress relieve certain alloy welds.

Q
QCP Quality Control Plan. A detailed method statement for repair work to be carried out in a turnaround

R
RACI Review, Action, Consult, Report: a responsibility assignment matrix
RAM Risk Assessment Matrix
RBI Risk-Based Inspection
RBMI Reliability-Based Mechanical Integrity
RCA Root Cause Analysis
RCM Reliability Centered Maintenance—a strategy that is implemented to optimize the maintenance program of a company or facility. The final result of an RCM program is the implementation of a specific maintenance strategy on each of the assets of the facility.
RFCA Root Failure Cause Analysis
RT Radiographic Testing—X-raying of welds

S
SAP German Enterprise Resource Management (ERM) software company
Scope Freeze Date Locking the project scope on or before the date on which the project baseline is set
SD Shutdown—planned or unplanned suspension of production
SHEQ Safety, Health, Environment, and Quality
SIMOPS Simultaneous Operations occur when two or more potentially conflicting activities are being executed in the same location at the same time
SOC State Oil Company
SOR Statement o Requirements
SOW Scope of Work
Stakeholders Individuals and organizations who are involved in, or may be affected by, project activities
Steering Committee Provides direction to the turnaround project manager and core team to ensure the turnaround meets the needs of the business
STOP A health, safety, and environment program to preempt unsafe acts whereby anyone is empowered to notify the guilty party and management by issuing an STOP card
SU Start-up
SV Schedule Variance

T

TA Turnaround

T&I Turnaround and Inspection

Task The Microsoft project term for activity

Time Now Activity progress reporting date and time. See also Data Date

TPM Total Productive Maintenance is a system of maintaining and improving the integrity of production and quality systems through the machines, equipment, processes, and employees that add business value to an organization

TRR Turnaround Readiness Review

Turnaround A shutdown that has been planned from more than a year in advance to ensure a run length in line with industry norms with minimal unplanned down days

U

UAUC Unsafe Act/Unsafe Condition

UAV Unmanned Aerial Vehicle—also called a drone

Utilization Actual production related to maximum sustainable capacity (measured as a percentage)

V

Variance (predictability) A measure of performance of a turnaround against the company's estimates and targets (short term)

W

WBS Work Breakdown Structure

WHB Waste Heat Boiler

WP&B Work Plan and Budget (see also BP&B)

X

X-raying See Radiographic Testing—used to view the integrity of a weld

Y

You The one to make it happen

Z

Zone An area of a process plant that includes a number of process units.

Appendix H: Consultants, online training and expert contractors

Consultants and trainers
General: turnaround project management

1. Shell Global Solutions. Available from: http://www.shell.com/business-customers/global-solutions.html
2. Asset Performance Networks. Available from: http://ap-networks.com/
3. Project Assurance Bobby Singh. Available from: http://www.projectassurance.com/
4. TA Cook TAR fighter Shutdown project simulation. Available from: https://www.tacook.com/en/expertise/turnarounds-shutdowns-outages/
5. Practical Management for Turnaround Professionals John McLay. Available from: https://www.pmtp.ca/
6. Petroskills North America. Available from: https://www.petroskills.com
7. Petroedge Asia. Available from: https://www.petroedgeasia.net/search/turnarounds

Asset integrity/management

8. Lloyd's Register Energy. Available from: http://rbi.lrenergy.org/about-lloyds-register-energy/
9. The Woodhouse Partnership Ltd. Available from: http://www.twpl.com/

Software

10. TenSix Consulting. Available from: https://tensix.com/about-ten-six/

Online training

11. Deltek Open Plan presentation. Available from: https://www.youtube.com/watch?v=nHNoujLiqjY
12. Primavera Demo. Available from: https://www.youtube.com/watch?v=zEL553w_snM
13. Enterprise APM v4 software On-Stage Demo. Available from: https://www.meridium.com/type/demo

Expert contractors
General

14. AMEC Foster Wheeler. Available from: https://www.amecfw.com/markets-and-services/markets/oil-gas-chemicals
15. SNC Lavalin Kentz. Available from: http://www.snclavalin.com/en/oil-gas

16. Fluor. Available from: http://www.fluor.com/services/maintenance/ams/turnarounds-outages
17. McDermott/Chicago Bridge and Iron Company. Available from: https://www.cbi.com/
18. Fouré Lagadec. Available from: www.fourelagadec.com
19. ABB. Available from: https://new.abb.com/oil-and-gas/services/by-service/engineering-and-consulting/maintenance-consulting/shutdown-turnaround-management

Gas turbines

20. GE. Available from: https://www.ge.com/power/services/gas-turbines

Compressors

21. BHGE Nuovo Pignone. Available from: https://www.bhge.com/axial-compressors

Distillation columns and separators

22. Sulzer Tower Field Services. Available from: http://www.sulzer.com/en/Products-and-Services/Tower-Field-Services

Temporary power

23. Aggreko Modular Power Units. Available from: https://www.aggreko.com

Catalyst handling
International
24. Catalyst Handling. Available from: http://www.catalysthandling.com/

North America
25. Mattawa. Available from: http://www.mattawaindustrialservices.com/services/catalyst-changeout-services-2/#&panel1-4

Middle East
26. Anabeeb. Available from: https://www.anabeeb.com

Postweld heat treatment

27. Stork (a Fluor Company). Available from: https://www.stork.com/en/products-services/mechanical-piping/heat-treatment-equipment

FCCU
FCC turnarounds
28. Ledwood UK. Available from: http://www.ledwood.co.uk/hpslider/cat-cracker-fccu-production-and-turnarounds/

29. MAN. Available from: https://dwe.man-es.com/energy-related-engineering/refinery-technology

30. Fouré Lagadec. Available from: www.fourelagadec.com

FCC slide valves

31. Tapco. Available from: http://www.tapcoenpro.com/fccu-slide-valves/

FCC cyclones

32. Van Tongeren. Available from: http://www.van-tongeren.com/fcc/

Heavy lifting

33. Mammoet. Available from: http://www.mammoet.com/

Appendix I: Capability assessment

Maximum score 400

Application	Score: Fully applied 4 Mostly applied 3 Partially applied 2 Not applied 1
0. Strategy	
1) Long-term turnaround plan (>5 years) 2) Turnaround itemized in annual budget 3) Turnaround framework documented 4) Use of computerized maintenance management as part of an enterprise resource management system 5) Use of asset performance management tools utilizing RBI, RCM, etc.	
1. Integration	
6) Turnaround charter with planned duration and OOM budget 7) Clear alignment of TA statement of commitment with the corporate vision and mission 8) Structured phased planning and execution 9) Using lessons learned for the next turnaround 10) Benchmarking and KPIs	
2. Scope	
11) Structured cross-functional scope screening undertaken 12) Scope challenge 13) WBS 14) Strict scope control process after scope freeze date	
3. Schedule	
15) Milestone schedule published and adhered to 16) Critical path modeling tools used to integrate all aspects of the execution phase 17) Models from previous turnarounds 18) Daily progress monitoring	
4. Cost	
19) Earned value method 20) Three budget estimates (OOM, prelim, and final) at different phases of development 21) Costs monitored daily during execution	

Continued

5. *Quality*	
22) Quality control plans used during execution 23) Quality records retained in database 24) Hold points adhered to during execution	
6. *Resources*	
25) Turnaround steering committee established 26) Core staff competent in both technical and management fields 27) Organization charts show clear delineation of roles and responsibilities 28) Resource leveling done in 2 stages: generic and specific 29) Productivity monitored during execution	
7. *Communication*	
30) Turnaround manual documented 31) Required meetings established 32) Logistics processes for support services including creation of a plot plan 33) Permit to work and access control strictly adhered to during execution 34) Work progress reporting carried out, progress analyzed, and corrective action taken on a daily basis during execution 35) Final turnaround report completed covering all aspects of the turnaround, and presented to the decision maker 36) Lessons learned documented for next turnaround 37) Turnaround models and maintenance and inspection data archived	
8. *Risk*	
38) Front end loading 39) Formal risk management 40) HSE plans 41) Phase audits carried out, especially at end of detailed planning phase 42) Risk mitigation carried out during execution	
9. *Procurement*	
43) Long-lead item ordering process implemented 44) Process applied for choosing and incentivizing the right experienced contractors 45) Contracting strategy optimized for high probability of achieving agreed duration 46) Adequate and competent client staffing to manage contractors 47) Contractor/vendor evaluations carried out	
10. *Stakeholder*	
48) Stakeholders identified 49) Stakeholders engaged 50) Stakeholders satisfied **Total**	

Total scoring results

<80 Poor or no application: help is required

80–160 Awareness of application: the learning process is just beginning

160–240 Understanding of application: progress is being made

240–320 Competent in application: almost there

320–400 Excellent in application: you are a pacesetter!

Appendix J: Author's comments

I have been involved with maintenance and engineering projects for my entire career. In the early days, as a refinery design engineer, I was baffled when the turnaround planners asked for the engineering construction turnover packages, in preparation for a turnaround, many months before the event. Only once when I became personally involved in planning turnarounds myself did I realize that the critical path models had to be built at a very early stage and then continually refined until the duration was within the range that had been set by top management in the previous year, when the proposal for the turnaround had been added to the business plan and budget.

My experience with building critical path models (CPM) started in the mid-1990s, when I was tasked with downloading all the network diagrams for a large refining complex, from Artemis (a mainframe CPM package) to an intranet-based package. At the time only two packages were available—Open Plan and PrimaVera. I chose Open Plan as I could download all the activities in comma delimited format and then edit each activity individually to get the models to function. It was a mammoth task since each process unit was a separate model. The models were successfully tested during a number of turnarounds. On my next assignment, I upgraded the Open Plan models from disk operated system to Windows, in preparation for a turnaround that included significant engineering work (see Box 5.15) the biggest turnaround ever for that refinery.

With regard to turnaround management experience, I have been fortunate in being involved in the replacement of fluid catalytic cracking cyclones, possibly the largest exercise that any refinery can attempt, in three separate refineries (see Boxes 4.9 and 6.6).

After certifying as a project management professional, I consulted and trained in the Middle East and South Africa for a number of years, which included training in turnaround management. This was where I learned that turnarounds are essentially complex projects and, to be successful, have to be managed as such. At the time there was very little literature available on turnaround management until Tom Lenahan produced the first book dedicated to the subject in 1999, followed by Bobby Singh in 2000. I am grateful for the advice both their books offer and have used both books for training and consulting, as well as for reference purposes for this book.

I joined Qatar Petroleum in 2002 where I gained further hands-on experience as a senior business analyst, senior auditor, and performance manager, until my final position of shareholder adviser before retirement.

I believe that my career has provided me with a broad range of skills and insights, and this book is a distillation of my experiences, as well as those of my review team and participants from my various training courses. My review team has been indispensable in ensuring that the book is current with regard to turnaround and asset management methodologies, and the application of suitable software. Richard McGrath, an ex-Chevron turnaround planning engineer, is currently a maintenance manager in a paper mill in New Zealand. Antonio Conti, an ex-Petronas offshore maintenance engineer, is now a mechanical engineer in a steel mill in Belgium. Current commissioning and start-up practices were reviewed by KhaiZen Foo, a chemical engineer involved with the start-up of the Petronas Refinery and

Petrochemical Integrated Development, part of the Petronas Integrated Petroleum Complex project in Jahor, Malaysia.

I have used the pronoun "he" throughout the text, purely for the sake of ease, since I do not wish to call attention to gender in situations where it's not relevant. However, I would like it to be known that I include all genders in such instances.

Comments are welcome. I can be contacted via e-mail at bhturnarounds@gmail.com.

Bruce

Index

385